CHEMISTRY RESEARCH AND APPLICATIONS

PYRENE

CHEMICAL PROPERTIES, BIOCHEMISTRY APPLICATIONS AND TOXIC EFFECTS

CHEMISTRY RESEARCH AND APPLICATIONS

Additional books in this series can be found on Nova's website under the Series tab.

Additional e-books in this series can be found on Nova's website under the e-book tab.

Chemistry Research and Applications

Pyrene

Chemical Properties, Biochemistry Applications and Toxic Effects

Petr Ruzicka
and
Tomas Kral
Editors

nova publishers
New York

For permission to use material from this book please contact us:
Telephone 631-231-7269; Fax 631-231-8175
Web Site: http://www.novapublishers.com

Library of Congress Cataloging-in-Publication Data

Pyrene : chemical properties, biochemistry applications, and toxic effects / [edited by] Petr Ruzicka and Tomas Kral.
pages cm
Includes bibliographical references and index.
ISBN 978-1-62417-291-5 (hardcover)
1. Pyrene (Chemical) 2. Biochemistry. I. Ruzicka, Petr, 1962- II. Kral, Tomas, 1968-
QP801.P639P97 2013
572--dc23

2012042097

Published by Nova Science Publishers, Inc. † New York

CONTENTS

PREFACE

Pyrene (Py) and various metabolites with a Py chromophore are powerful probes often used in biophysical and biochemical sciences due to their high fluorescent quantum yields and relatively long fluorescence lifetimes. In this book, the authors present current research in the study of the chemical properties, biochemistry applications and toxic effects of pyrene. Topics include the biological effects induced by benzypyrene in soil and plants; benzo(a)pyrene and colonic cancer; bacterial metabolism of pyrene; low-temperature laser-induced fluorescence studies of various biological systems with pyrene chromophore; photo-energy and electron donating ability of pyrene; and synthesis of novel silylated pyrene units and their application as a fluorescent labeling agent for bioactive components such as oligodna and cholesterol.

Chapter 1 - Biological effects induced by benzpyrene (BP) in the soil-plant system have been studied. Transformation of the molecular composition of polycyclic aromatic hydrocarbons (PAH) of soils in the process of contamination of soil by benzpyrene (BP) and growing *Tradescantia* (clone 02) has been detected. The effect of selective formation of PAH in soils predetermined by the amount of the BP applied has been identified. Application of small BP dozes causes the formation of mostly light 3,4-nuclear polyarenes, higher BP dozes form heavy 5,6-nuclear PAHs. After integrating BP into soils in small doses, products of its metabolism produce auxin effect on *Tradescantia* (clone 02) and barley (*Hordeum vulgare L.*).

Morphological deviations induced by BP have been detected in flowers and in filament fuzzes of *Tradescantia* (clone 02). On the basis of frequency of display of rosy (RMD) and white mutant developments (WMD) the assessment of BP mutant impact on a plant was made. Various adaptation levels of *Tradescantia* (clone 02) to BP influence was detected. On the morphological level the adaptation of the plant is not expressed, adaptation can be observed for physiological effects – white mutant developments and mitosis inhibition (MI). At the genetic variation level in plants all stages of adaptation according to Selye's stress theory are exhibited. When studying the genetic and toxic effect of oil on *Tradescantia* (clone 02) similar regularities were observed.

Total toxicological reactivity of polycyclic aromatic hydrocarbons has been calculated. Major share of the toxicological reactivity of PAH is constituted by 5,6-nuclear polyarenes, with BP making a significant proportion, its mass fraction in soils and plants was minimal though.

Chapter 2 - Colorectal cancer (CRC) is one of the most common cancers in the Western world. Subsets of food-derived heterocyclic amines and *N*-nitroso compounds are probably the environmental chemicals that cause CRCs in humans. However, the contribution of environmentally–ubiquitous polyaromatic hydrocarbons (PAHs) toward causing human CRCs is still unclear. One of the probable reasons for this is that not so much attention has been paid to the colonic carcinogenicity of PAHs. The number of PAHs that have been examined is extremely limited for colonic carcinogenicity induced by oral treatment (by gavage or in the diet). The most representative PAH that is carcinogenic is benzo[*a*]pyrene (BP). The carcinogenicity of BP has not been noted in the colon of rodents, although forestomach, spleen, lung, and liver are shown to be target organs for the carcinogenesis of BP by oral treatment. However, the authors showed that BP was mutagenic in multiple organs of mice, including the colon with the highest mutation frequency. Recently, the authors demonstrated that BP can rapidly induce colonic adenocarcinomas in mice at a high incidence/multiplicity in the presence of dextran sulfate sodium, which is non-genotoxic and induces colitis in the murine colon. Furthermore, the authors examined the mutagenicity of BP with the Ames test, a routine *in vitro* mutagenicity test used world–wide, using multiple human liver S9 fractions in addition to liver S9 fractions from rats pretreated with inducers of drug–metabolism enzymes. BP was weakly or marginally mutagenic in the majority of human S9 fractions; however, some of them showed potent mutagenicity with individually diverse mutagenic activities. These findings may partially account for the induction of colonic tumors by BP in humans, particularly to patients with inflammatory bowel disease. Thus, further studies of the carcinogenic risk of PAHs including BP on the colon are strongly recommended.

Chapter 3 - Pyrene is a peri-condensed four-ring polycyclic aromatic hydrocarbon (PAH). It is highly recalcitrant and a model compound for the study of degradative pathways of other PAHs such as benzo(a)pyrene. *Mycobacterium* species and some other actinomycetes were the first pyrene metabolizing bacteria to be described starting from 1988. Today, pyrene is known to be metabolized by a broad range of bacteria including Gram-negatives such as *Pseudomonas, Sphingomonas, Stenotromorpha, Alcaligenes, Achromobacter, Leclercia, Cycloclasticus and Burkholderia* among others. The main pathway of pyrene metabolism is via dioxygenation at the 4,5-position followed by *ortho* cleavage to produce 4,5-dihydroxy-4,5-dihydro phenanthrene and subsequent sequential cleavage leading to the tricarboxylic acid (TCA) cycle via O-phthalate. Another pathway, less common, proceeds through the cleavage of the two central rings at the 4, 5-and 9, 10-positions to produce a unique metabolite 6,6'-dihydroxy-2,2'-biphenyl dicarboxylic acid which also proceeds through O-phthalate but in fewer steps. Dioxygenation at the 1,2 position often leads to methylated dead-end products or detoxification. There are also reports of cometabolism of pyrene in the presence of co-substrates such as phenanthrene and fluoranthene. Bacterial metabolism of pyrene is modulated by factors such as pH, temperature, salinity, biosurfactants and humic acids amongst others. Products of incomplete metabolism of pyrene have been shown to inhibit degradation of other PAHs in the environment. In *Mycobacterium* species, especially, genes for pyrene metabolism are localized both on chromosomes and on plasmids. However, not much is known yet about the metabolism and genomics of Gram-negative pyrene degraders. Therefore, efforts are expected to be directed in this direction in the near future.

Chapter 4 - Pyrene (Py) and various metabolites with a Py chromophore are powerful probes often used in biophysical and biochemical sciences due to their high fluorescent quantum yields and relatively long fluorescence lifetimes. Their fluorescent properties are very sensitive to the change in their immediate surroundings and are used to probe their local environment. In this chapter, the authors review recent applications using low- and high-resolution fluorescence-based methodologies (i.e., the non-line-narrowing (NLN) and fluorescence line-narrowing (FLN) spectroscopies) that utilize their electronic properties. These methodologies are used in conformational studies of biomolecules containing Py chromophores and in identification of enantiomeric metabolites containing Py moiety. The authors demonstrate this approach allows probing the ground- and excited-state vibrational frequencies of various metabolites and DNA adducts, and can provide insight into the binding site to DNA bases in various *in vitro* and *in vivo* experiments. The position of the fluorescent origin-band and the so called "fingerprint" of molecule(s) based on their characteristic vibrational frequencies can also provide insight on the mechanism of DNA adduct formation. It is also shown that laser-based fluorescence spectroscopies can successfully probe interactions of metabolites with monoclonal antibodies, including cross-reactivity and conformational multiplicity.

Chapter 5 - Photoinduced electron transfer and energy transfer can be possible in the photoexcited pyrene to porphyrin. In the molecular system, which consists of the pyrene and the porphyrin, the electron transfer from the photoexcited pyrene to the ground state porphyrin moiety competes with the energy transfer from the pyrene to the porphyrin. These photophysical processes could be controlled by the solvent polarity and the distance between the pyrene and the porphyrin in the pyrene-phosphorus(V) porphyrin system.

Using a similar molecular system, these photophysical processes can be selected by the pH of the solvent. In the case of the direct connecting molecular system of the pyrene and the porphyrin, ultra-fast electron transfer was observed. Furthermore, the electron transition process of pyrene may be used for the control of the singlet oxygen generation photosensitized by the porphyrin moiety in a water-soluble pyrenylporphyrin.

Chapter 6 - The silylation of pyrene causes increments of molar extinction coefficient and fluorescence quantum yield in addition to a bathochromic effect (approximately 10 nm) inboth absorption and emission maxima. In order to take advantage of these unique features, the authors of this study developed several novel silylated pyrene analogs and some bioactive components, such as oligoDNA and cholesterol, labeled with silylated pyrene.

Experimental results show that the silylated pyrene can be utilized as a sensitive fluorescent labeling unit for bioactive materials. A summary of the findings is found in this chapter.

1) The silylation of pyrene was achieved *via* the lithiation of bromopyrene, followed by the reaction with appropriate chlorodimethylsilyl agents possessing a modifiable functional group. Subsequent to silylation, a suitable combination of chemical reactions afforded silylated pyrene derivatives with terminal OH or NH_2 groups. The OH group was then converted to phosphoramidite function which is readily applicable to modern DNA/RNA synthesizers utilizing solid-phase phosphoramidite chemistry.
2) The silylated pyrene phosphoramidite was used to label the 5'-terminus of oligoDNA *via* phosphodiester linkage. The resulting oligoDNA exhibited a faint fluorescent signal so long as the oligomer stayed in a single strand. The signal, however, intensified markedly in the presence of a fully complementary (non-

mismatched) DNA strand. Interestingly, one-base mismatch introduction into the complementary DNA quenched the signal intensification. Meanwhile, the silylated pyrene containing the NH_2 group was utilized to label the C2'-position of uridine *via* carbamate linkage. The modified uridine was further derived to the corresponding 3'-*O*-phosphoramidite which was subsequently incorporated into two consecutive positions of oligoDNA. This oligoDNA exhibited a marked excimer fluorescence signal only upon binding to a fully complementary DNA molecule. That signal was, in turn, quenched into a single-stranded form and in a duplex with a mismatched base pair. As can be seen, the silylated pyrene-modified oligoDNAs demonstrated a sequence-discriminating property characterized by a lightening of its fluorescence signal.

3) The authors also applied the silylated pyrene unit to visualize intracellular cholesterol traffic in macrophages. Thus, silylated pyrene possessing NH_2 group was reacted with a cholenic acid derivative. The resulting silylated pyrene-cholesterol exhibited an enhanced fluorescence quantum yield compared to that of unmodified pyrene in methanol. In living cells, silylated pyrene-cholesterol moves to oil droplets in macrophages yet it remains in the ER membrane due to the acyl-CoA:cholesterol acyltransferase (ACAT) inhibitor (Sandoz 58-035). These results seem to indicate that cholesterol which contains silylated pyrene is very suitable for the visualization of intracellular cholesterol traffic due to its enhanced fluorescence signal and behaves similarly to natural cholesterol.

In: Pyrene
Editors: Petr Ruzicka and Tomas Kral

ISBN: 978-1-62417-291-5

Chapter 1

Biological Effects Induced by Benzpyrene in Soil and Plants

Evgenia V. Yakovleva,[1] Dmitriy N. Gabov,[1] Vasiliy A. Beznosikov,[1] Boris M.Kondratenok[1] and Maria V. Aniskina[2]

[1]Institute of Biology of Komi Scientific Centre of the Ural Branch of the Russian Academy of Sciences, Syktyvkar, the Komi Republic, Russia
[2]Syktyvkar State University, Syktyvkar, the Komi Republic, Russia

Abstract

Biological effects induced by benzpyrene (BP) in the soil-plant system have been studied. Transformation of the molecular composition of polycyclic aromatic hydrocarbons (PAH) of soils in the process of contamination of soil by benzpyrene (BP) and growing *Tradescantia* (clone 02) has been detected. The effect of selective formation of PAH in soils predetermined by the amount of the BP applied has been identified. Application of small BP dozes causes the formation of mostly light 3,4-nuclear polyarenes, higher BP dozes form heavy 5,6-nuclear PAHs. After integrating BP into soils in small doses, products of its metabolism produce auxin effect on *Tradescantia* (clone 02) and barley (*Hordeum vulgare L.*).

Morphological deviations induced by BP have been detected in flowers and in filament fuzzes of *Tradescantia* (clone 02). On the basis of frequency of display of rosy (RMD) and white mutant developments (WMD) the assessment of BP mutant impact on a plant was made. Various adaptation levels of *Tradescantia* (clone 02) to BP influence was detected. On the morphological level the adaptation of the plant is not expressed, adaptation can be observed for physiological effects – white mutant developments and mitosis inhibition (MI). At the genetic variation level in plants all stages of adaptation according to Selye's stress theory are exhibited. When studying the genetic and toxic effect of oil on *Tradescantia* (clone 02) similar regularities were observed.

Total toxicological reactivity of polycyclic aromatic hydrocarbons has been calculated. Major share of the toxicological reactivity of PAH is constituted by 5,6-

nuclear polyarenes, with BP making a significant proportion, its mass fraction in soils and plants was minimal though.

Introduction

Oil and gas industry is one of the most environmentally hazardous branches of our economy. Being a universal and remarkably convenient source of energy, oil at the same time is a potential threat for the environment and is one of the main pollution factors (Volikov, 1989; Danilov, 1996).

The process of oil processing generates sludge as waste, it is a compound mixture of oil hydrocarbons, water and suspended organic and nonorganic substances. The content of hydrocarbon part in oilsludge reaches 10-65 %. As a result of oil processing with each 1 ton of oil up to 14 kg of sludge are generated. Storage of oilsludge in storage bunkers is connected with the pollution of underground waters and atmosphere (Sorokin, 1983). Lately, oil contamination of soils as a result of industrial accidents during oil extraction and transportation has been acquiring more and more significance (Panov et al., 1986). It is known that oil is one of the most hazardous contaminants due to its properties and scale of its use.

From the chemical point of view oil presents a combination of various in chemical structure hydrocarbons with the admixture of their derivatives, including mostly sulphur, oxygen and nitrogen, as well as asphaltenes and resins. Oil hydrocarbons are to a bigger extent represented by compounds of three classes: paraffines (alkanes) with the formula C_nH_{2n+2}, cycloparaffines (naphthenes) with the formula C_nH_{2n} and aromatic hydrocarbons with the formula C_nH_n (Barenboym et al., 2000). When getting into the soil crude oil containing up to 50 % of light (with the boiling temperature up to 250°C) fractions is subject to physical and chemical influence: evaporation, elution, ultraviolet irradiation, as a result of which light water-soluble hydrocarbons, decompose and migrate away from the spill. Remaining heavy bituminous oil product is exposed to biochemical degradation. The decisive power lies on microorganisms performing intracellular oxidation of hydrocarbons.

Oil and oil products migrate downwards across the soil body. During the vertical movement across the soil body a chromatographic effect is created which leads to the differentiation of the composition of oil and oil products. In the upper humus horizon high-molecular components containing resin-asphaltene substances and polycyclic compounds get accumulated. Low-molecular compounds with high water solubility mostly get to lower horizons. Acids of high viscosity oils, apart from carboxylic acids, also contain phenols, cresols and their derivatives (Goncharov, 1987). The most critical biospheric function of phenols is fixation of toxic substances, such as heavy metals, pesticides, radioactive elements, which leads to the decrease of toxicity of fixed substances and the change of their behaviour (Elin, 1997).

After oil or oil product spills two processes start taking place on the soil surface simultaneously – evaporation of hydrocarbons into the atmosphere and their filtration into the soil. Evaporation rate is predetermined by physic-chemical properties of oil and oil products and the ambient temperature. Intensity oil hydrocarbons penetration in the soil body will depend on the soil structure and properties of oil and oil products, first of all on viscosity and the content of high-molecular resin-asphaltene substances (Filchenkova, 1982).

Oil, oil products, coals, combustion gases used as chemical material and fuel for industrial companies and transport serve as main sources for the formation of polycyclic aromatic hydrocarbons PAH (Shabad et al., 1979; Rubtsova, 1997).

PAH is a compound having cancerogenic, mutagenic and other toxic impact on living organisms (Ilnitsky, 1975; Nielsen, 1996; Bispo et al., 1999). It is assumed that cancerogenic PAH cause tumors mostly at the contact location, but some tumors can appear in distant organs as well. Cancerogenic PAH, widespread both at industrial sites and in the environment, influence the incidence of cancer, first of all, of lungs and stomach. Aromatic hydrocarbons are the most toxic oil components. With the weight fraction of PAH in water 1 % all water plants disappear. Mutagenic reactivity of PAH is similar to cancerogenic one. For example, 3,4-benzpyrene is referred to the class of the most powerful mutagens. It is established, for instance, that its influence on DNA leads to the replacement of nucleotides. PAH in bitumens is known to cause genotoxic impact. Genotox cancerogens do not have a threshold level, i.e. their presence in any amount is hazardous for a living organism (Danetskaya,1971).

Unsubstituted single hydrocarbon – benzpyrene – is most frequently defined as an indicator of cancerogenic PAHs in natural objects. This hydrocarbon together with other heavy PAHs (dibenzpyrenes, dibenzanthracene, benzfluoranthenes) relates to first class of hazard. As a rule in objects containing benzpyrene other PAHs can be identified as well (Shabad, 1979; Gennadiev et al., 1996; Belykh, 2005).

PAHs are higher priority pullutants both in the list of the European Community (EC), and in the list of the US Environmental Protection Agency (ERA) (Jian et al., 2004). Out of 49 PAH compounds assessed within the framework of the international expertise 10 compounds were defined as potentially harmful for animals and people. Among them are benzanthracene, benzfluoranthene, benzpyrene, dibenzanthracene, indene[1,2,3-cd]pyrene and others (Nadala et al., 2004). PAHs have extremely low maximum allowable concentrations (MAC). At the moment in Russia there are MAC for benzpyrene, pyrene, naphthalene, phenanthrene in the working area air and for water bodies (Gennadiev et al., 1996; Shabad.,1979). In Russia only benzpyrene is subject to mandatory monitoring in soils. Benzpyrene MAC in soils is 20 ng/g; in drinking water – 1 ng/l (SN 2.1.7.2041-06).

Practically in all natural and industrial objects PAHs can undergo various chemical modifications: biological degradation and photochemical destruction. Polyarene content in ecosystems is in dynamic equilibrium (Gennadiev et al., 1996; Shabad, 1979).

The most effective decomposition of PAHs can be observed in acidic soils (Tonkopy, Shestopalova et al., 1979). In the soil with RH = 4.5 during the first 10 days from 95 to 99 % of deposited benzpyrene is degraded, whereas in soils with RH = 7.2 – from 18 to 80 %. Later, the decomposition rate slows down significantly. For soil, which is a self-regulating non-living system, self-regenerating processes are typical. It is known that soil microorganisms can destroy PAHs (Shabad, 1979). Apart from microbiological and chemical modifications other factors like metabolism in plants and invertebrates, fermentation activity, water resistance, gaseous interchange and sorbing properties of soils can play a significant role in the process of self-purification of soils. Self-purification of soils depends on the chemical composition of polluting agents including PAH, and in southern regions is completed faster than in northern ones (Rovinsky et al., 1988).

In soils when UV radiation or powerful chemical oxidizers are absent, biological systems are the most influential agents in the processes of PAH decomposition. The ability of living organisms to recycle PAH by means of including them into their metabolism was observed with many aerobes – animals, plants, microorganisms. It is established that PAHs are decomposed in tissue culture, including human fibroblast. Hundreds of species of bacteria, fungi, higher plants which use PAH as carbonaceous substrate for generation of free energy have been identified (Shurubor, 2000; Shabad, 1979).

The study of the process of PAH biological degradation showed that the most efficient method of their destruction is microbiological oxidation. (Dikun, 1979). Microbiological degradation of PAHs is exceptionally selective in character with regards to single compounds. The selective character of PAH oxidation and rate of the process are prominently influenced by original chemical properties of single PAHs and their composition: number of aromatic rings in a molecule, presence and size of pendent groups, the degree of saturation of rings.

It is known that soil microorganisms can destroy as many PAHs as they "are used to" (Shabad, 1972). For example, if the soil is heavily polluted and contains a lot of PAHs, its microbial flora is able to destroy up to 70-80 % of their original amount. A sharp increase of the PAH content in the soil leads to the decrease of weight of metabolizing structures.

Low-molecular PAHs are characterized by lower chemical reactivity as compared to PAHs with higher condensation. However, cultures of various microorganisms can modify efficiently and rapidly only low-molecular PAHs, whereas high-molecular PAHs oxidized insignificantly or failed to oxidize (Mc Kenna, 1976). It is suggested that this process partially depends on permeability of cell membranes, and is also determined by the biological reactivity of oxidized compounds. In its natural state the cell membrane is partially permeable: water flows through it almost without obstacles, the diffusion rate of other substances is directly proportional to their solubility and inversely proportional to their molecular weight. Weak biological oxidability of high-molecular PAH can be explained by their insignificant solubility. As a result of increased solubility low-molecular PAHs have an easier access into the cell (Shurubor, 2000). Physiologically active compounds (vitamins, hormones, etc.) in their molecular structure have 2-3-nuclear structures of naphthalene and phenanthrene types. Natural similarity with such substances can to some extent influence as a stimulus on permeability of cell membranes.

For a big number of microorganisms hydrocarbons serve as the source of carbon and energy (Boonchan et al., 2000). Experiments have shown that for 11 months microorganisms metabolized up to 86 % of high-molecular PAHs (from chrysene to benz[ghi]perilene) and up to 19 % low-molecular PAHs (from acenaphthene to pyrene) of the initial amount (Zheng and Obbard, 2002). Even at low temperatures in northern and arctic soils in 90 days aerobiocultures modified from 52 to 88 % of PAHs (Eriksson et al., 2003).

Flora also participates in accumulation and processing of hydrocarbon compounds. Along with the ability for intracellular degradation of PAHs, higher plants and algae in most cases demonstrate the tendency to accumulate hydrocarbons from their habitat without any visual structural changes of hydrocarbon skeleton (Dikun, 1979). It was proved by the scientists that after fixing organic compounds can move from roots to other plant organs and later undergo full or partial decomposition or transform into less toxic compounds and bind in plant tissues. Organic substances deposit in cell vacuoles or bind with insoluble cell structures, such as lignin (Salt at al., 2007).

Thus, chemical and biological PAH transformation processes in natural environment possess varied directions complimenting each other and ensuring at the end a unified process of biogeochemical transformation of PAHs in ecosystems.

It has become more frequently in recent days that biological systems are used for indication of pollutants in the environment. The reaction of an organism allows to assess the environmental impact in indicators which are of biological meaning. When influencing the environment, physiological factors and chemical compounds are sometimes significantly transformed by factors of animate and inanimate nature; their full impact cannot be predicted easily. Bioindicators create an accurate and complete picture including those pollutants that were missed by a monitoring service (Bioindication of terrestrial ecosystem pollution, 1988)

The soil-plant system is an important object for the monitoring of environmental pollution with PAH. Soils and vegetative communities function in close cooperation with each other. Notwithstanding this fact, the peculiarities of PAH accumulation in plants and soils are studied insufficiently. The role of polluted soil in the accumulation of PAH by plants is still debatable. Along with the data about polyarenes absorption from soils by plants (Raykhman, 1999), there are data about absence of such absorption (Tonkopy et al., 1979). Some scientists believe that the ability to accumulate PAH is a species character of plants (Simonich, Hites, 1994).

S.V. Durmishidze et al. (1979) in their studies have established that plant leaves and roots uptake benzpyrene and transform it into various compounds. Most of all, benzpyrene is localized in mitochondria and plastids – organelles playing the most important role in oxidizing. High doses of benzpyrene break the structure of these cell organs: mitochondria reduce, their cells get partially destroyed, the form of the nucleolus changes, and the nucleus lightens. It was proved that wheat and rye germs grown in different soils and on nutrient liquid absorbed hydrocarbon with roots, when benzpyrene was added there. The benzpyrene uptaken with roots was transported to above-ground organs unchanged (Ugrekhelidze, 1979).

At present the absorption of PAH from soil and atmosphere has mostly been studied for cultivated plants, regarding wild plants the data are not numerous. The data on the absorption of PAH are ambiguous. A number of researches have estimated that PAH due to its high lipophilicity and low solubility is absorbed by root epidermis when contacting with soil particles, but is not uptaken into the root (Kipopoulou et al., 1999; Backer et al., 2000; Tao et al., 2004). Other authors state that polyarenes get into plant tissues mostly from soil than from air (Fismes et al., 2002).

Experiments with growing a number of agricultural croppers on the soil which is polluted with pyrene and phenanthrene showed the possibility of decomposition of them by plants (Gao Yan-Zheng et al., 2005; Xu et al., 2006). The researchers note that the decomposition of PAH by plants decreased as the PAH molecular weight increased (Smith et al., 2006). It is established that with high benzpyrene doses in soil, benzpyrene precipitates on the root surface and the number of zones of absorption becomes limited (Durmishidze et al., 1979; Galiulin et al., 2005; Xiao-Dong Huang et al., 2004).

The research devoted to the distribution of PAH in plant organs demonstrated that the accumulation of polyarenes in organs depends on the species of a plant. Oats incepted benzpyrene most aggressively by the growing rhizosphere of a stem (Chernichenko, Pavlova, 1979). The content of benzpyrene in wheat reaches its maximum in heads, less in roots, then in leaves, and in stems the content of benzpyrene is at its minimum (Tonkopy et al., 1979). German scientists found in tropical trees organs 21 PAHs. Their research showed that the proportion of polyarenes goes down along the line leaves > stems > branches > bark. (Krauss et al., 2005).

In the literature the data can be found about the ability of plants to synthesize PAH. In nutrient solutions deprived of carbonaceous substrate benzpyrene synthesis was not observed. During seed sprouting in solutions containing carbonaceous substrate (with tracer carbon), benzpyrene synthesis up to 7 mkg/kg of seeds was observed (Dikun, Kalinina, 1979; Shabad, Khesina, 1979).

The behavior of pollutants in the system soil-plant builds up of a several processes: absorption by the plant roots, remission, migration (desalination) and degradation in soil. So, according to the concept of foreign authors the absorption of pollutants from the soil by roots is seen as the function of substance solubility in water, of organic matter content and plant species (Simonich, Hites, 1995).

The detailed analysis of accumulation of resistant organic matter by plants allowed to suggest that the accumulation coefficients (proportion of matter content in roots to the one in soil) linearly depend on their content in soils. Estimations show that in general the transition of resistant organic matter upwards from soil to plants is higher (35 – 70%) than to water (12 – 18%) or atmosphere (18%), (Gaponyuk, 1977).

An important indicator for determination of the nature of accumulation of benzpyrene by plants from soil is benzpyrene solubility in water and its ability to transit from various substrates to water. Studies demonstrated that from sand to water extract benzpyrene transfers in such quantities that exceed its solubility significantly. The increase of a benzpyrene dose deposited into sand leads to the increase of its content in wheat organs. On clayey soils PAH accumulation in plants is less than on sabulous soils which is connected with the increased absorbing ability of soils formed on argillaceous solids (Pavlova, Donina, 1979). The level of ecosystem pollution is mostly defined by low-molecular PAH, polyarenes of average weight make an insignificant contribution. High-molecular PAH (indenopyrene, dibenzO[a,h]anthracene benz[g,h,i]perilene) stay in soils and plants in trace quantity. (Meudec et al., 2006).

With the help of the research about the content of benzpyrene, fluoranthene and benz[g,h,i]perilene in the leaves of city landscape plants it was shown that most pollutants were detected in plants growing in the city area (Slepyan et al., 1979). During comprehensive environmental research of a number of areas of the southern Baikal region it was established that a significant accumulation of PAH in the vegetation cover was found near an aluminum plant. Regarding the content of PAH botanical objects were classified in the following order: grass sod > miscellaneous herbs of nonarable territories > dead grass > top of potato and carrot > birch leaves > permanent grasses > oats (Belykh et al., 2004; Belykh 2005).

Research by national and foreign authors demonstrated an excellent state of knowledge about the character, accumulation and migration of PAH in soils. However, the systematic approach to the exploration of polyarene accumulation in the environment is insufficient. The study of PAH is to a great extent fractionary and is aimed at the analysis of distinct ecosystem components. The characteristics of PAH accumulation in the soil-plant system are insufficiently explored. When studying the pollution of environmental objects with PAH mostly pollution with benzpyrene was analyzed, other polyarenes included into priority pollutant lists are poorly explored.

The objective of the research: to estimate the genotoxic effect of oil hydrocarbons and benzpyrene on plants and identify regularities in bioaccumulation and transformation of polycyclic aromatic hydrocarbons in the soil–plants system on the example of *Tradescantia* (clone 02).

Objects and Methods of Research

Objects of Research

As objects of research we used highly sensitive plants of *Tradescantia* (clone 02). Tradescantia belongs to the family of *Commelinaceae R.Br.* Many species of the *Tradescantia* genus grow in the natural conditions of North America. Many of these species or their interspecies hybrids were collected and grown in the Brookhaven National laboratory (US).

Clone 02, which is most frequently used for the assessment of mutagenic impact of chemical and physical factors (Grant et al., 1994; Ma, 1997), was taken from the natural population. It presents a hybrid between *T. occidentales Pritton* ex. Rydb. and *T.* ohiensis Raf., and it is heterozygous regarding the colouring of the flower (Osipova, Shevchenko, 1984). The height of plants of such a clon varies from 30 to 60 cm. Each inflorescence consists of several florets of different development stages. In normal conditions one flower of an inflorescence blooms every day (Figure 1).

In a flower there are three sepals, three petals, six stamens and one pistil. Well-developed stamens have 30-60 hairs, each of each contains up to 32 cells (these figures may change depending on the growing conditions). A hair is a set of meristemic cells consecutively placed and developing from one epidermal cell by means of fragmentation of terminal and subterminal cells (Underbrink et al., 1973; Evseyeva, Geraskin, 2001).

Hair cells of staminal threads in florets with most active cell fragmentation are the most sensitive field for mutations. The changes in the gradation of cells (bends, branching) and in their sizes are known to be morphological abnormality (MA). The appearance of morphologically abnormal cells can be connected not only with mutations, but with distortion of physiological processes as well. And although there is data about induction of such distortions in hair cells of staminal threads by various mutagens (Mericle, Mericle, 1967), it is normally not used as a criterion for the estimation of mutagenic impact.

The blue colouring of the flower cells is prepotent, the rosy one - recessive. The transformation of the blue colouring of the flower cells of staminal threads into the rosy one is considered to be a phenomic mutation in cells which are heterozygous in colouring.

Figure 1. The structure of spiderwort (*Tradescantia*) flower.

As a result of fragmentation of mutant cells in a mature hair a mutant sector including one or more cells is formed. Somatic mutations can be calculated both in stamen hairs and in petals. In our research only stamen hairs were used.

Due to the influence of various factors the loss of cell reproductive power can be observed. To determine the hairs suspended in growth it is necessary to determine the number of cells in a hair. The inhibition of cell fragmentation can occur in any hair along the length of the stamen. The average length of a hair in the top third of a stamen is approximately 16 cells, whereas hairs in the middle and at the bottom can be much longer. Most hairs of tradescantia (clone 02) in the top stamen part consist of 12 and more cells. Therefore such hairs are considered normal, while hairs with less cells are defined as the ones having lost their reproductive power (Underbrink et al., 1973). This allows to analyze the survival ability of cells after stress effect.

A number of studies on radiation and chemical mutagenesis have established that somatic cells of staminal threads of tradescantia are similar in their sensitivity to cells of animate organisms regarding the loss of reproductive power, and according to the criterion of "frequency of somatic mutations" even outreach them (Mericle, Mericle, 1967; Underbrink et al., 1973; Dennis, 1976). Genetic variability of tradescantia is minimal, as the plants are mostly reproduced by vegetative propagation (by axillary cuts). But their rising generation which was cultivated by means of pollinating is highly mutable. According to different authors, spontaneous frequency of mutations for clone 02 varies between 0.048 and 0.15 % (Sparrow et al., 1977; Ichikawa, Sparrow, 1978).

Tradescantia (clone 02) is a convenient object for studying the impact of low doses of various factors. From this point of view its advantages as a test-object include its high sensibility to external effects at low spontaneous mutation level, low individual variability, accuracy and simplicity of external impact marker registration. The possibility to use this test-object and the reliability of data about environmental pollution received with its help are confirmed officially by many leading laboratories of the world (Ma et al., 1997).

Soils of the model experiment. In model lab tests topsoil (Arable layer 0÷20 cm) of the cultivated podzolic formed on silt loam was used. Morphological characteristics of the soil is given on the example of the transverse section 5 km to the south-west from Syktyvkar at the watershed divide of the Sysola and Vazhel-u rivers. The relief is plain with low hills. Soil-forming material is medium-textured coarse clayey soil, at the depth of 140 cm there are carbonaceous encrusting matters. Exploration period – about 30 years.

A_{top} 0-30 cm – sandy clay loam, grey, peat flakes, small concretions, loosely cloddy, humid, with a lot of roots, with white spots of podzolic horizon and reddish brown spots of illuvial horizon. Level clear contact.

A_2B_1 30-47 cm – sandy clay, light-brown with grey spots, with laminal nuciform structure with white powder. The upper part of the horizon is compacted (plough pan), with thin roots, dark leaks, a lot of aeruginous spots, irregular contact.

B_1 47-82 cm – silty clay, reddiosh brown with coarse nuciform structure, white powder at glassy facets of structural joints, sporadic roots, with hardpan grains, water leaks over cavities, gradational contact.

B_2 82-120 cm – sandy clay of the brown colour, coarse nuciform structure, small hardpans, sporadic thin plant roots, gradational contact.

BC 120-140 cm – sandy clay of the brown colour, non-structured, viscous, aeruginous spots.

Table 1. Agrochemical parameters of cultivated podzolic soil

Layer	Sampling depth, cm	pH, salt	Acidity		Adsorbed cations		ω (N total), %	Humus content, %	Labile forms		The rate of ion saturation, Ca^{+2} and Mg^{+2}, %
			Hydro-lytic	exchangeable	v (Ca^{+2})	v (Mg^{+2})			ω (P_2O_5)	ω (K_2O)	
			mmol/ 100 g of soil						mg / 100 g of soil		
A_{top}	0-10	6.5	1.0	0.02	9.4	5.8	0.09	2.8	103	34	94
A_{top}	10-20	6.6	0.8	0.01	9.2	5.8	0.06	2.4	93	43	96
A_{top}	20-30	6.6	0.8	0.01	9.2	5.1	0.07	2.9	156	48	95
A_2B_1	30-40	5.3	1.2	0.01	7.1	6.4	0.02	0.5	7	42	84
A_2B_1	40-50	4.3	2.9	0.99	8.8	8.0	0.00	0.5	6	41	86
B_1	50-60	4.2	3.3	1.09	11.4	9.5	0.03	0.5	9	24	94
B_1	60-70	4.2	2.9	0.80	13.1	9.9	0.02	0.5	14	17	89
B_1	70-80	4.5	2.0	0.37	13.9	9.9	0.03	0.3	23	12	92
B_2	80-90	4.8	1.4	0.07	13.5	9.5	0.04	0.2	28	11	94
B_2	90-100	5.2	1.2	0.04	13.8	8.9	0.03	0.2	30	10	95
B_2	100-110	5.5	1.0	0.03	13.0	9.0	0.02	0.2	33	9	96
B_2	110-120	5.7	0.8	0.03	13.4	9.1	0.03	0.2	35	8	96
BC	120-130	6.3	0.6	0.02	12.8	8.5	0.03	0.4	37	9	97
BC	130-140	7.3	0.2	0.00	26.5	8.6	0.01	0.3	23	16	99

Table 2. Parameters of oil from Verchnevozeyskoe production field

Density at 20°C, t/m ³	Viscosity at 20° C, mPa	Water content, % volume	Content, % mass		
			resins	silicagel resins	asphaltens
0.82-0.85	6-12	4.9	0.25	6-6.7	0.3-1.1

$Cg`_{Ca}$ 140-150 cm – sandy clay of the brown colour, at lower part there are blue-grey spots, non-structured, viscous, aeruginous spots, boils from hydrochloric acid.

Agrochemical properties of cultivated podzolic soil are provided in table 1.

Soil of arable horizon was put into glass containers, the soil weight in each container was 400 g. Benz[a]pyrene was added to soil in the form of water solution with the appropriate concentration prepared by means of diluting the original state standard sample (SSS) with the benz[a]pyrene content 200 mg/cm^3 (SSS № 7515/98). The amount of added benzpyrene complied with its content in containers 10, 20, 30 and 40 ng/g of soil. Nil treatment – without benz[A]pyrene.

Replications in tests – 6 times. The tests were conducted in laboratory conditions with the room indoor temperature (20-25°C) and light. The soil humidity in containers was sustained with the help of distilled water at the level of 60 %.

Plants were grown during a month, then they were cut and their total plants biomass was calculated, as well as separately roots and leaves biomass was assessed. Estimation of PAH enleaching by corresponding plant organs and by whole plants was performed for air-dry weight. In the identical way soil was polluted with oil from the Verkhnevozeyskoye field of Usinsk district (table 2). The content of oil in soil was 0; 0.25; 0.5; 1; 1.9; 3.8; 7.5 and 15 %. Plants were grown in equal soil conditions, treatment varied in the dosage of oil pollution in soil only.

Methods of Research

Study of Genotoxicity of Oil and Benzpyrene for Tradescantia (Clone 02)

As flowers appeared, starting from the 4th and till the 50th day after application of benzpyrene, hair cells of staminal threads of tradescantia were daily observed on the subject of somatic mutations with the criterion of prepotent blue colouring transformed into the recessive rosy one. In addition, the account of morphological abnormalities was conducted: the amount of giant and dwarf cells, bends and branching. White mutant developments and cell fragmentation inhibition were also accounted separately.

Before the analysis, staminal threads were separated from the base by a pincer with great care and placed into a drop of vegetable oil on the microscope glass, then hairs were straightened carefully, and the objects were studied under the microscope with magnification of 15×8.

The frequency of MA of rosy (RMD) I white mutant developments (WMD) was calculated as the proportion between the amount of mutant cells and total amount of hair cells in staminal threads. Single cells or a group of mutant cells separated by one or more normal cells were considered as the result of two mutant developments. The indicator «inhibition of cell fragmentation» was calculated as a share of hairs which failed to survive (less than 12 cells) from the total amount of cells under study. In total, 39442 hair cells of staminal threads (HSF) were analyzed in the lab experiment.

Determination of PAH in soils and plants. PAH I soils was conducted according to the method PND F 16.1:2:2. 2:3. 39-03., certified in accordance with GOST R 8.563, – «Methods of measurement of mass fraction of benz(a)pyrene in samples of soils, subgrades, bed deposits, and solid wastes by means of HPLC, with the use of liquid analyzer «Fluorat 02» as a "fluorimetric detector". The error of PAH measurement in soils for confidence coefficient P = 0.95 makes 35 % for the measurement range from 5 to 40 ng/g inclusively and 25 % for the range from 40 to 2000 ng/g inclusively. The basis of PAH determination in plants is the original method of extracting hydrocarbon components (Yakovleva et al., 2008). The method of PAH determination was tested wirth the soil standard «PAH contaminated soil/sediment» (Resource Technology Corporation, US). The following PAH were identified: phenanthrene, anthracene, fluoranthene, pyrene, benz [A] anthracene, chrysene, benz [b] fluoranthene, benz [K] fluoranthene, benz [A] pyrene, dibenz [a,h] anthracene, benz [ghi] perilene, indeno [1,2,3-cd] pyrene. Standard deviation was from 3 to 25 % of the measured values of the component content (program Statistica 6.0. «Statsoft»). In order to estimate quality of chemical analysis of PAH we used "Organochlorine compounds and petroleum hydrocarbons in seaweed (*Fucus sp.)*" for plants and "Standard reference material 1944 New York – New Jersey Waterway sediment" for soil.

Results and Discussion

Genotoxic Effect of Oil on Tradescantia (Clone 02)

In the study of genotoxicity *Tradescantia* (clone 02) has the leading role (Ma et al, 1997; Grant W.F. et all, 1994; Knasmuller S et al, 1998; Evseyeva et al., 2000). Today it is almost the only test-system fit for the identification of mutagenity of atmospheric pollutants in situ and it is recommended as an express-method at the first stage in the system of genetic monitoring of environmental pollution (Schairer et al, 1978; Ichikawa, 1991). The visible marker used in this test-object is a phenotypic transformation of pigmentation from blue to rosy in HSF cells. The monitoring of somatic mutations in HSF turned out to be quite a simple and therefore attractive method of study mutagenesis under the effect of a wide range of mutagenic factors. After the contamination of soil with oil different deviations were found in the development of HSF *of Tradescantia* (clone 02). Even at the morphological level the response of the vegetative system of stress was observed. The frequency of exposure of morphological abnormalities in HSF of changed nonlinearly with the peak at the dose of 0.5 % (Figure 2). Changes in the frequency of display of MA are mostly preconditioned by giant cells (GC). In the similar way the frequency of display of MI changed (Figure 3) The appearance of giant cells is suggested to be connected with the blocking of mitosis and loss of control over the The appearance of giant cells is connected with blocking the mitosis and the loss of control over growth and extension.

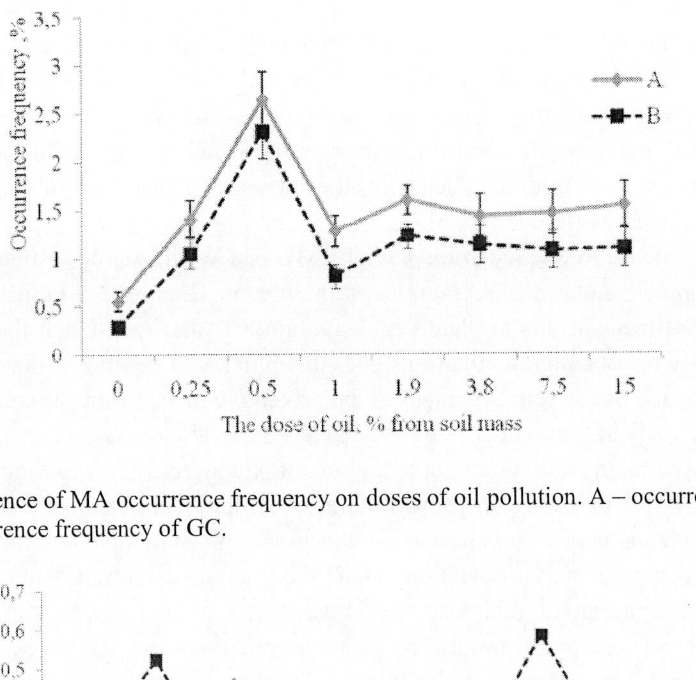

Figure 2. Dependence of MA occurrence frequency on doses of oil pollution. A – occurrence frequency of MA, B – occurrence frequency of GC.

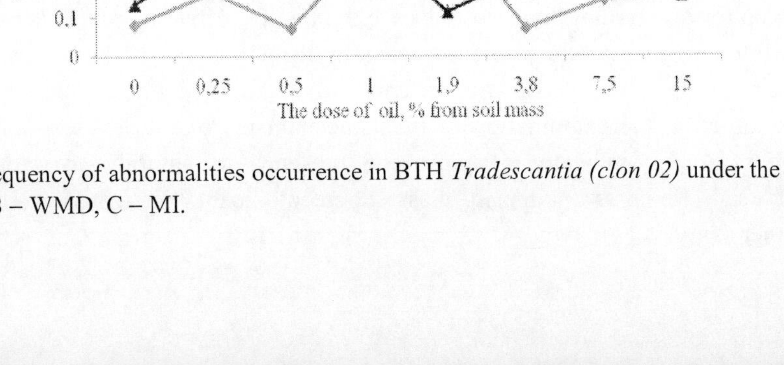

Figure 3. Frequency of abnormalities occurrence in BTH *Tradescantia (clon 02)* under the oil pollution: A – RMD, B – WMD, C – MI.

It is probable that the increase in frequency of MI display in such a case stimulates the appearance of giant cells at the dosage of 0,5 % oils in soil.

In the course of the experiment the increase of the display of WMD and RMD was observed after the pollution of the soil, as compared to control. WMD, as well as MI are morphological deviations. The appearance of white cells is connected with inhibiting of pigment development in plants. The indicator white mutant developments varies significantly depending on the pollutant doses. First reactions of plants on the stress are expressed in the increase of the display of WMD at the dose of oil in the soil at 0,25 %. With the dose of 0,5 % all mechanisms of adaptation of plants to stress effects switch on. The malfunction of adaptation mechanisms starts emerging at the dose of 7.5 % oil in soil. Further decrease of the frequency of display of WMD can be connected with the decrease of number of flowers in plants at the maximum pollutant dose.

Nonlinear dependence of low radiation dose effect has recently been studied. It was demonstrated that low-scale radiation can produce a more mutagenic effect than some radiation with high doses (Zaynullin, 1998). For oil pollution such dependence was shown for the first time.

The emergence of rosy cells is predetermined by heterozygosity of plants regarding the colouring of the flower. In the normal state the stamen have prepotent blue colouring which is determined by a genetic factor (D+) which is responsible for the development of delphinidin, the pigment. As a result of stress impact (in this case the effect of oil hydrocarbons) dominant allele shuts down and gives place to the recessive one which is responsible for the synthesis of the rosy pigment – cyanidine. Therefore, the increase of the level of somatic mutations indicates the mutagenic influence of oil.

The changes in the frequency of display of RMD and WMD are opposite. The increase of the level of somatic mutations in samples with low oil dosage can indicate the fact that adaptive response mechanisms in plant cells have not activated yet. Upon the increase of oil dosage the frequency of somatic mutations goes down in some samples till the reference level which can be caused by the reinforcement of the protective mechanisms of cells.

The analysis of the deviation frequency in the hairs of *Tradescantia* staminal threads (clone 02) showed mutagenic, teratogenic and genotoxic effect of oil pollution of soil on the plant, with this effect displayed in proved increase of the level of somatic mutations, white mutant developments, cell fragmentation inhibition and morphological abnormalities. It was detected that the appearance of deviations under study depends nonlinearly on the degree of pollution. The adaptation of plants to the stress factor started displaying already on the morphological level. The peak of morphological abnormalities and MI belonged to the dosage of 0,5 % oil in soil. The most prominent reaction of the plant to stress was shown as WMD and RMD. On the basis of the data obtained it should be noted that the examination of the oil effect mechanisms in plants is of great interest. Oil is a multicomponent system, and it is hard to estimate the influence of its various fractions on plants. At high dosage, the toxicity of oil can be preserved for a very long time. In this case the biggest effect on plants is produced by polyaromatic fractions and products of intermediate decomposition of heavy hydrocarbons and heterocyclic compounds. PAHs are the most toxic oil components, therefore further research were aimed at the examination of plant mechanisms of PAH absorption and their genotoxic effect. As an indicator of polyarene presence in natural ecosystems mostly benzpyrene is used, which is a 1 hazard class superecotoxicant. So, it was selected as the object for further study.

Genotoxic Effects in Tradescantia (Clone 02) Induced by Benzpyrene

The issue of genetic safety has been acquiring more and more significance in connection with the increased environmental pollution and the growth of cancer rate. The range of chemicals and physical factors influencing a human or other organisms, including PAH which is an indicator of all aforementioned, is expanding every year.

Due to its high biological activity and cancerogenic, mutagenic and teratogenic effect, benz [a] pyrene is subject to mandatory control in environmental objects. For the reason of the existence of various PAH sources leading to the pollution of the soil-plant system, there is a threat for genetic safety for the majority of the population.

As a result of injection of various doses of benzpyrene into soil, *Tradescantia (clone 02)* displayed a number of morphological deviations (Figure 4). The frequency of display of morphological deviations in *Tradescantia (clone 02)* flowers steadily grew with the increase of the dose of benzpyrene injected into soil. In the sample without benzpyrene morphological deviations were observed in 0.12 % of flowers under study.

With the dose of benzpyrene of 40 ng/g the frequency of display of morphological deviations reached 1,74 %. The increase of frequency of appearance of flowers with unexpanded flower buds, coalescent sepals and petals was observed. In the no-treatment sample and at the pollution with the 10 ng/g dose coalescent petals were absent.

In the same was the frequency of display of MA in hair cells of *Tradescantia (clone 02)* staminal threads changed. The increase of the MA display was reflected through more frequent appearance of bends in *Tradescantia (clone 02)* HSF with the increase of pollution (table 3.).

The dosage dependence of the frequency of display of morphological deviations among *Tradescantia (clone 02)* flowers and HSF was logarithmical and represented an initial stage of classical S-curve.

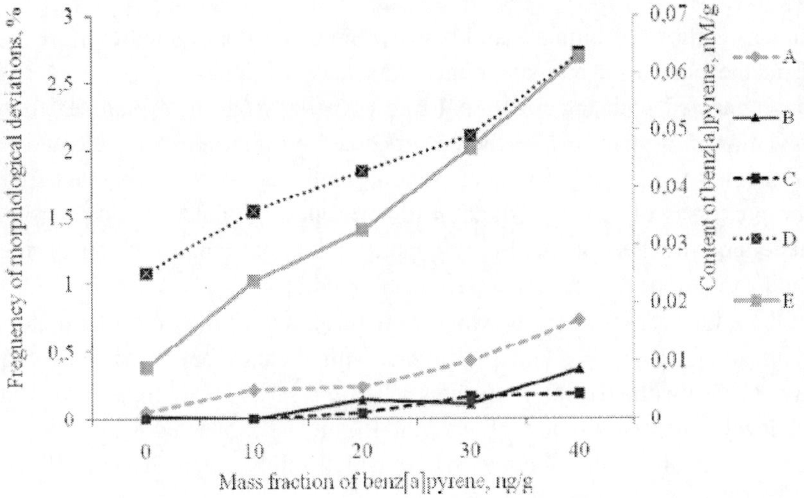

Figure 4. Frequency of *Tradescantia (clon 02)* flowers morphological deviations occurrence. A – unopened buds, B – accreted sepals, C – accreted petals, D – frequency of HSF curves, E – content of benz[a]pyrene in plant, nM/g.

Table 3. Level of morphological abnormalities of filaments hairs of *Tradescantia* (*clon 02*), %

Mass fraction of benz[a]pyrene, ng/g of soil	Change of the size of cages				Change of an arrangement of cages		Total MA	
	GC		DC		Bend			
	$\bar{X}$	$S\bar{x}$	$\bar{X}$	$S\bar{x}$	$\bar{X}$	$S\bar{x}$	$\bar{X}$	$S\bar{x}$
0	0.08	0.04	0.30	0.06	1.08	0.12	1.58	0.15
10	0.20	0.09	0.75	0.29	1.54	0.24	2.59	0.43
20	0.36	0.13	0.70	0.13	1.83	0.24	3.06	0.35
30	0.30	0.09	0.72	0.18	2.10	0.23	3.28	0.26
40	0.25	0.11	0.71	0.18	2.71	0.31	3.81	0.40

The display of morphological deviations is connected with the teratogenic effect of benzpyrene, as its content in plants increased in the similar way as MA. The data obtained prove that adaptation mechanisms of plants were not observed on morphological level at the given benzpyrene pollution rate. As a result of oil pollution of soil adaptation was already observable at the morphologicl level which was predetermined by the frequency of GC display.

If the soil is polluted with benzpyrene, giant and dwarf cells (DC) (developments connected with the modification of the size) reacted differently on different doses of benzpyrene, than MA. The frequency of display of dwarf cells after the pollution was reliably higher than reference values and changed slightly at different pollution doses.

The GC frequency curve ($R = 0.93$) was identical to the plant biomass transformation curve with the peak at the dose 20 ng/g of benzpyrene (Figure 5). It seems that the decrease of GC frequency at high pollutant doses was connected with the decrease of the total plant biomass.

Curvilinear relations of biomass and benzpyrene doses determined to a greater extent by the change in the plant root biomass rather than leaves biomass. It is possible that such a change was connected with the functional role of benzpyrene in a plant. Benzpyrene after transcarboxylation in a plant or after being processed by microorganisms could perform the function of auxin, the phytogormone, (β-indoleacetic acid), and (or) benzpyrene could intensify the processes of auxin production in aerial part of plants. Even microquantities of auxin regulate the growth of plants. In big quantity it can be found in plants during blooming stage and bud-formation. Auxin main effect is targeted at the growth of root system and cello extension (GC) – these are the effects which were observed during the experiment.

An identical reaction on soil pollution with benzpyrene was received by T.V. Krestovskaya (1979) with co-authors in roots of *Allium sera.* The increase of auxin dose over the optimal level causes inhibition of growth. It can be connected with auxin stimulating synthesis of another phytohormone – ethylene which inhibits plant growth (Polevoy, 1986; Polevoy et al., 1991; Yakushkina et al., 2005).

The experiment of growing barley (*Hordeum vulgare L.*) with the pollution of the substrate with benzpyrene doses of 10, 20, 30 and 40 ng/g of soil showed that the biomass of aerial part of barley decreased linearly following pollution, and the biomass of roots had nonlinear dependence with the peak at benzpyrene rate in soil 20-30 ng/g.

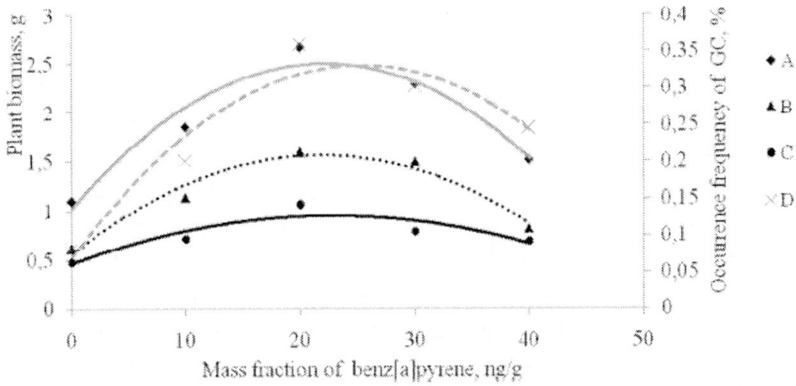

Figure 5. Dependence of HA occurrence frequency and tradescantia biomass of added doses of benz[a]pyrene. A – whole plant biomass, B – roots biomass, C – leaves bimass, D – GC occurrence frequency.

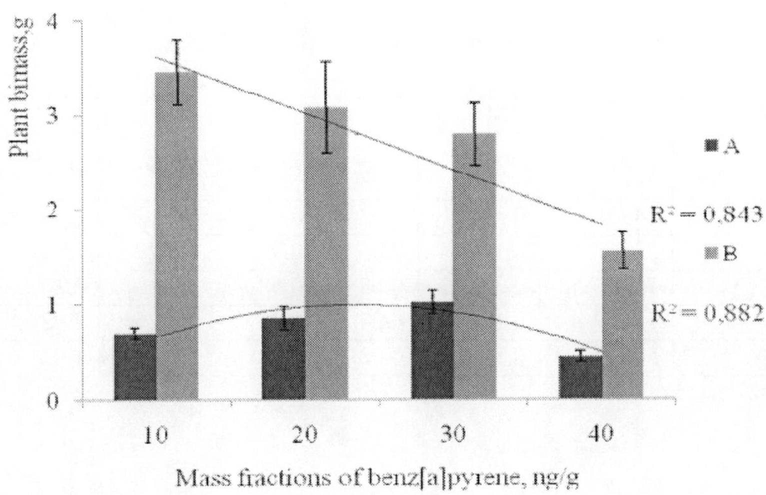

Figure 6. Dependence of barley biomass of the level of benz[a]pyrene soil pollution A – roots biomass, B – biomass of the aerial part.

All that agrees with the data obtained on tradescantia and supports auxin effect of benzpyrene in plants (Figure 6). It should be noted that following oil pollution the frequency of display of GC was higher than following benzpyrene pollution. It is possibly explained by the presence of a large quantity of substances with auxin-like structure in oil. Biological oxidation of aromatic hydrocarbons in cells is performed as a result of free-radical processes which cause the appearance of arene-oxides capable for covalent bonding with nucleophilic cell fragments (proteins, nucleic acids, etc.) which activate peroxidation of lipids in biological membranes.

As a result, mutagenic, cancerogenic, and cytotoxic influence of polyarenes is initiated (Skugoreva et al., 2008; Kutsenko, 2002).

The course of HSF examination demonstrated the tendency to the increase of WMD and MI in *Tradescantia* (clone 02) following the increase of pollution rate by benzpyrene (table

4). WMD and MI are referred to morphological deviations with their display predetermined by various mechanisms, such as the disruption of the cell cycle, hormone rearrangement, etc.

The proved increase of the quantity of inhibited hairs following the pollution rate was detected. Frequency of display of deviations in cell inhibition acts is practically twice higher for all polluted samples as compared to the reference sample.

Cell fragmentation inhibition is explained by the delay of mitosis which signifies explicit toxic effect of benzpyrene in plants. The decrease of fragmentation rate and cell extension under the influence of benzpyrene can arise from various mechanisms: direct linking to DNA, prolongation of mitotic cycle, disturbance of microtubule formation, decrease of cell membrane plasticity, reduction of glutathione pool (Skugoreva et al., 2008).

Identically to MI, the frequency of display of WMD changed (Figure 7). The indicator of white mutant developments varied from 0.68 to 0.98 %, whereas in reference samples at benzpyrene doses of 10, 30, 40 ng/g this indicator was reliably higher than the reference value (with R= 0.95). Regarding the frequency of cell fragmentation inhibition and the display of white mutant developments decline was observed with the dose of 20 ng/g benzpyrene in soil.

Table 4. Level of deviations in filaments hairs of *Tradescantia* (clon 02)

Mass fraction of benz[a]pyrene, ng/g of soil	Total HSF	RMD, %		WMD,%		MI, %	
		$\bar{X}$	$S\bar{x}$	$\bar{X}$	$S\bar{x}$	$\bar{X}$	$S\bar{x}$
0	14735	0.11	0.03	0.75	0.12	1.27	0.12
10	5316	0.11	0.05	0.97	0.22	2.14	0.31
20	7824	0.27	0.08	0.68	0.15	2.10	0.26
30	6566	0.10	0.04	1.15	0.25	2.92	0.42
40	5001	0.30	0.08	0.98	0.17	1.93	0.26

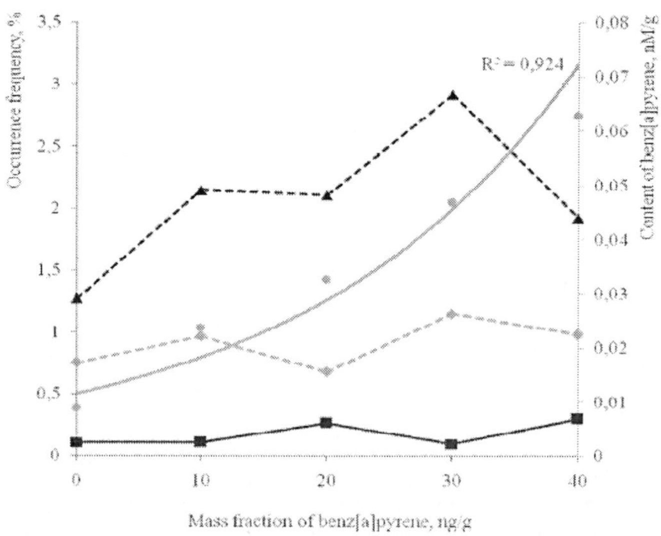

Figure 7. Occurrence frequency of RMD, WMD and MI in HSF of *Tradescantia (clon 02)* flowers. A – WMD, B – MI, C – RMD, D – content of benz[a]pyrene in plant, nM/g.

It might be probable that with such pollution rate the plant acquires resistance to such mutations showing mechanisms of cell reparation.

In the experiment the sample dosed with 30 ng/g of benzpyrene in soil is characterized as having the highest toxicity expressed according to the quantity of WMD and MI. The plant loses the ability to adapt, and further decrease of frequency of WMD and MI is only caused by the reduction of plant biomass and plant flowers. Similar regularities were obtained after pollution of soil with oil. The increase of the amount of rosy and white mutant developments proves mutagenic effect of benzpyrene on plants. The display of rosy mutant developments is explained by heterozygosity of plants with regard to the flower colouring. RMD are true genetic disruptions, they appear when the prepotent blue colouring gene is eliminated.

The frequency of rosy mutant developments transformed from 0.11 to 0.30 %. In samples with benzpyrene mass fraction of 20 and 40 ng/g the rate of these deviations reached its maximum, which is contrary to the changes in the frequency of display of WMD and MI. Frequency of display of RMD is the most sensitive indicator of plant reaction on stress effect. On the basis of the conducted research three intervals of dose dependence of benzpyrene accumulation by a plant and frequency of display of somatic mutations in hair cells of *Tradescantia (clone 02)* staminal threads were identified (Figure 8).

In the first interval of the curve – the interval of small doses of benzpyrene in soil (0 – 20 ng/g) the typical increase of the benzpyrene content in a plant was observed, as well as of the frequency of display of RMD. The plant was in the phase of apprehension and showed the primary reaction on stress – the plant biomass grew and the total amount of PAH in a plant increased, PAH content was prevailed by compounds with trinuclear molecule structure (phenanthrene).

In the second interval – the range of average doses (20 – 30 ng/g) you could observe the sharp decline of RMD due to further increase of benzpyrene content in plants. The decrease of the frequency of display of RMD at such benzpyrene dose can be explained either by the triggering of reparation mechanisms in plants, or by cell apoptosis which can appear in the form of MI.

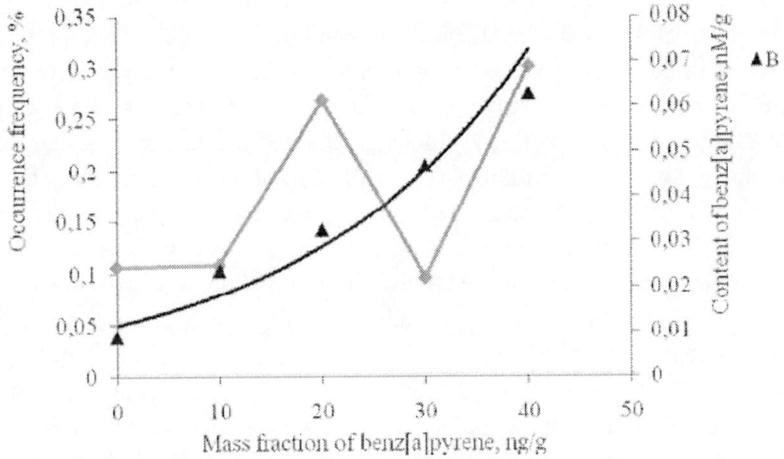

Figure 8. Curves of benz[a]pyrene accumulation in plant and frequencies of mutations occurrences. A – RMD frequency, B – content of benz[a]pyrene in plant, nM/g.

At that, the plant biomass practically remained the same, the total PAH content in plants went down, and the share of heavy polyarenes in the total PAH of plants increased.

In the third range – the area of high benzpyrene doses (30 – 40 ng/g) – intensive RMD growth was observed which was connected with the sharp increase of benzpyrene content in plants. Such a phenomenon can be explained by the disruption in plants of stress adaptation mechanisms. The increase of the frequency of display of RMD following soil pollution with 40 ng/g of benzpyrene was accompanied by the decrease of plant biomass (Figure 5.), which later could lead to the demise of *Tradescantia (clone 02).* With the decrease of the biomass the PAH content in plants also declined. The proportion of heavy PAH in the total amount of PAH in plants became more and more significant. In general the reaction of plants to soil pollution with benzpyrene had curvilinear relation which is characteristic of all biological systems and corresponded well to the Selye stress theory (Selye, 1977). In this case the stress factor was the presence of a powerful toxicant – benzpyrene – in soil.

Thus, we can observe 3 types of reaction of *Tradescantia (clone 02)* to benzpyrene. On the morphological level the reaction was expressed in the form of an ascending curve identical to the curve of benzpyrene growth in a plant. The reaction on the cell level (GC) is expressed by the curve with the peak at the dose of 20 ng/g, and is identical to the plant biomass curve ($r = 0.90$, with $n=6$, $R= 0.95$) – which are the consequences of auxin effect of benzpyrene. On these levels plant adaptation to high doses of benzpyrene in soil is not displayed. RMD, WMD and MI frequency curves are nonlinear, WMD and MI are completely contradictory to RMD frequency curve, the reaction of plants in this case lags behind as the indicators of WMD and MI are less sensitive, than RMD. On the other hand, the increase of MI following the decline of RMD level can be explained by the phenomenon of cell apoptosis.

Similar results were received for *Tradescantia (clone 02)* frown in soil polluted with soil. It should be noted that the frequency of display of all deviations as a result of soil pollution with benzpyrene is 10-20 times higher than after the pollution with oil. The exception is the frequency of display of GC which is higher at pollution with oil.

One of theories that helps to explain the toxic effect of excessive doses of oil and benzpyrene is the stress concept. The effect of excessive doses of pollutants triggers plants response in the form of the oxidative stress. As a result of stress the generation of active form of oxygen (AFO) in the vegetative cell increases to such extent that protective mechanisms are overdone inducing the "oxidative burst" effect. In high doses AFO cause various oxidative transformations in a cell, they damage nucleic acids, proteins, suspend the cell cycle, induse apoptosis transformations. AFO trigger reactions of lipid peroxidation leading to the disturbance of cell membranes. Vegetative organisms are resistant to oxidative damage. It is explained by the effective antioxidising system in a vegetative cell which includes antioxidative ferments and organic antioxidants. Antioxidant properties are also characteristic for compounds decreasing the reactivity of radical oxidizing processes, such as glutathione, ascorbic acid, tocopherol, thyroxin, steroids, ferritins, carotinoid, retinol, polyamines, ubiquinone (Selye, 1977; Merzlyak, 1989; Minibaeva, 2003; Skugoreva et al., 2008)

The research conducted helped to detect the dose dependence of the frequency of display of morphological deviations in flowers and HSF of *Tradescantia (clone 02)*, the level of physiological disruptions in HSF cells, the frequency of mutations on various doses of the injected benzpyrene. It was established that benzpyren in examined doses caused a clear toxic effect in plants. The response of plants to different benzpyrene doses was ambiguous and was expressed to a greater extent in various mutagenic, teratogenic and genotoxic effects. Frequency of display of morphological deviations and MA among *Tradescantia (clone 02)* increased with the growth of pollution rate. The peak of WMD and MI display was observed at benzpyrene doses of 10 and 30 ng/g.

The most sensitive indicator of the plant reaction to stress caused by the injection of benzpyrene in soil was the frequency of display of RMD in hair cells of *Tradescantia (clone 02)* staminal threads, which reflected all stages of plant adaptation: primary stress, active adaptation, disruption of adaptation processes followed by the death of an organism (Yakovleva et al., 2011).

Thus, the obtained data on the influence of oil and benzApyrene pollution of soils on the frequency of mutant developments in hair cells of *Tradescantia (clone 02)* staminal threads signified the display of mutagenic properties of oil and benzpyrene, the increased level of morphological abnormalities in hair cells of staminal threads – teratogenic effect, and the increased display of cell fragmentation inhibition – their genotoxicity. The dose dependence of PAH accumulation by the plant and the mutation rate was expressed in a classic S–curve which reflects different stages of adaptation.

Biological Effects Induced by Benzpyrene in Soil and Tradescantia (Clone 02) Plants

The importance of knowledge of fundamental processes of PAH redistribution in the soil-plant system, quantitative characteristics of these processes, identification of regularities of biological effects of polyarene impact on the system are the basis for the development of recommendations for the preparation of safe polyarene-protectors. Well-known research on bioaccumulation of PAH in soils and plants do not provide any idea about the regularities of bioaccumulation of toxicants by plants, its transformation in soils, and the quantitative indicators of these processes are practically missing.

Few results of the definition of biological absorption of PAH in the soil-plant system are widely variable and do not allow to establish factors influencing its values. The examination of interrelation of the exchange of polycyclic aromatic hydrocarbons and soil can provide a comprehensive and correct vision on the impact of soil pollution on the life of plants, as well as reliable theoretical and practical grounds of the methods of rehabilitation of soil medium from various pollutants.

Alteration of Molecular Content of PAH in Soils under the Influence of Benzpyrene

The molecular content of PAH of the original soil sample is shown in pic. 2. In arable podzolic soil 8 single PAHs were detected (Figure 9). Weight fraction of PAH with 3,4-nuclear structure in the original soil constituted 70 % (phenanthrene, pyrene, fluoranthene, chrysene) of their total content. Among heavy PAH with 5,6-nuclear structure benz[*b*]fluoranthene prevailed. Naphthalene, anthracene, fluorene, benz[a]anthracene, benz[ghi]perilene were not found in soil.

It was established that in the process of growth and development of plants in the sample without benzpyrene (reference sample) in soil the reduction of PAH content decrease in soil went on, to a greater extent, due to chrysene and benz[b]fluoranthene.

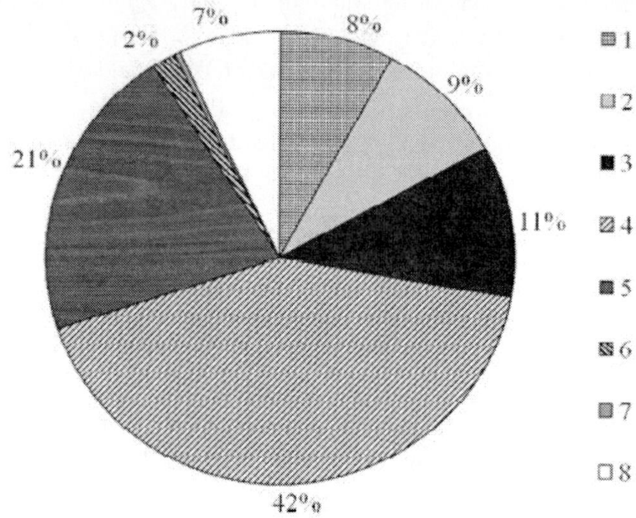

Figure 9. Mass fraction of identified compounds in total PAH sum in original soil: 1 – phenanthrene, 2 – fluoranthene, 3 – pyrene, 4 – chrysene, 5 – benz[b]fluoranthene, 6 – benz[k]fluoranthene, 7 – benz[a]pyrene, 8 – dibenz [a,h]anthracene.

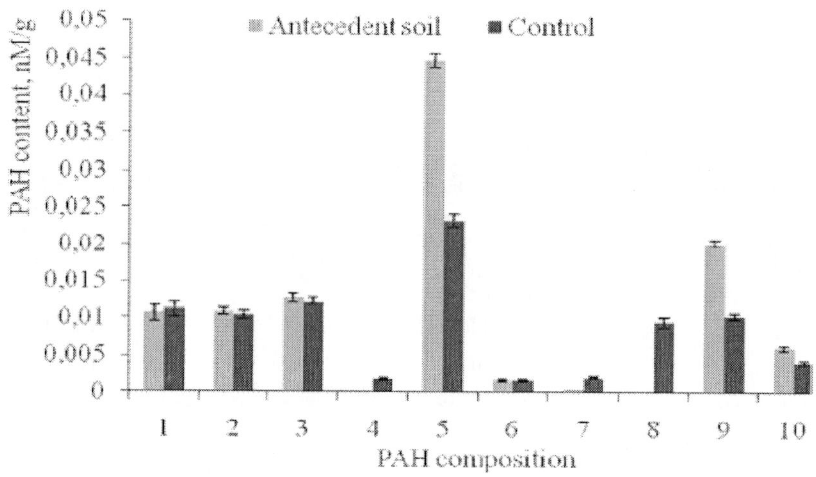

Figure 10. Content of PAH in original soil and in soil of control variant after the experiment: 1 – phenanthrene, 2 – fluoranthene, 3 – pyrene, 4 – benz[a]anthracene, 5 – chrysene, 6 – benz[k]fluoranthene, 7 – benz[a]pyrene, 8 – benz[ghi]perylene, 9 – benz[b]fluoranthene, 10 – dibenz[a, h,]anthracene.

The present fact allows to speak about the biological role of PAH for a vegetative organism. The share of light polyarenes in the total amount of PAH decrease in the reference sample soil by the end of the experiment made 95 % (Figure 10).

In the reference sample the display of benz[ghi]perylene in significant quantity was detected, which was not found in the original soil. At that, the increment of benz[ghi]perylene compensated the loss of benz[b]fluoranthene.

After soil contamination with benzpyrene in different doses changes in the content of some polyarenes were detected (Figure 11).

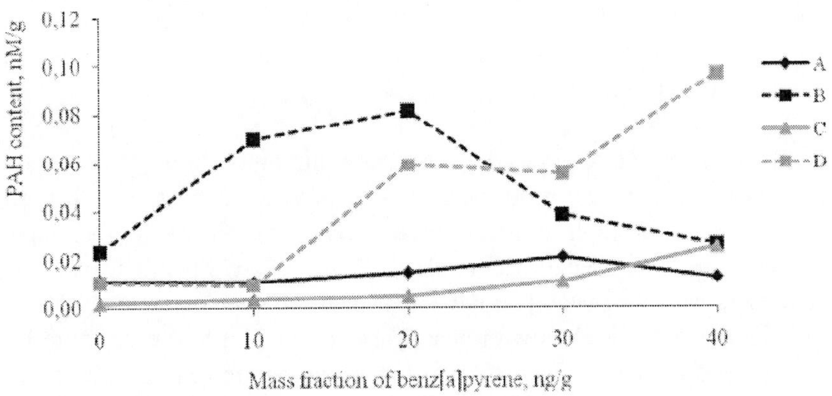

Figure 11. Content of PAH in soil: A – phenanthrene, B – chrysene, C – benz[a]pyrene, D – benz[b]fluoranthene.

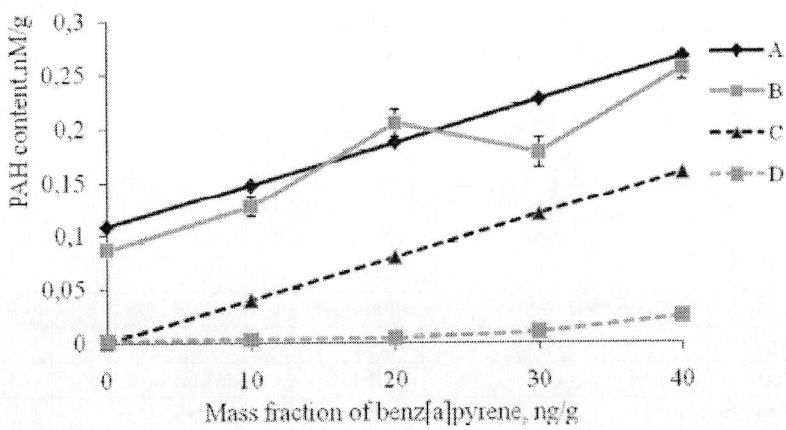

Figure 12. Content of PAH in soil: A – phenanthrene, B – chrysene, C – benz[a]pyrene, D – benz[b]fluoranthene.

With the pollution level of 10 – 20 ng/g a significantly increased content of chrysene in soil was observed, and with higher doses of benzpyrene – its sharp decrease. The proportion of heavy PAH: benz [b] fluoranthene, benz [k] fluoranthene grew with higher pollution rate. The highest growth was observed for benz [b] fluoranthene. The changes in the content of other PAH are insignificant.

The comparison of PAH content in soil after the injection of benzpyrene and after the experiment allowed to establish that the total content of polyarenes in soil for all experiment samples remained practically unchanged (Figure 12). However, the content of benzpyrene in soil after the experiment plummeted sharply.

The transformation of the molecular composition of polycyclic aromatic hydrocarbons in soil is connected with high reactivity of benzpyrene, which lead to the selective formation of "extra - PAH" – an additional amount of mobilizable polyarenes from soil as a result of the effect of the injected benzpyrene (table 5, Figure13). Lighte PAH formed after the injection of low doses of benzpyrene into soil, high – as a result of its high weight fraction. The proportion of heavy PAH: benzO [k] fluoranthene, benzpyrene, benzO [b] fluoranthene, benz

[ghi] perylene among “extra-PAH” went up with the increase of benzpyrene dose injected into soil. As a result of soil pollution with the dose of 10 ng/g, the increment of PAH was performed by chrysene and benz[ghi]perylene, which compensated for the loss of benzpyrene.

The increased content of “extra-PAH” with the pollution of 20 ng/g was conditioned by the increase of chrysene and benz[b]fluoranthene in soil, which completely substituted for the loss of benzpyrene. With high doses of benzpyrene in soil its losses stopped being compensated by chrysene, benz[b]fluoranthene and benz[ghi]perylene, the growth of benz[k]fluoranthene was observed. The increment of these compounds with doses of 20-30 ng/g remained on the same level as with benzpyrene doses of 10-20 ng/g. With benzpyrene dose of 30 ng/g the total increase of PAH was lower than the reduction of benzpyreneA, i.e. benzpyrene either decomposed to non-aromatic components, or absorbed by a plant. Maximal content of “extra-PAH” was observed with the soil pollution with 40 ng/g of benzpyrene, the increment of PAH in such a sample practically completely compensated for the loss of benzpyrene. Transformation processes in soil lead to consecutive decrease of the proportion of light hydrocarbons in the total content of PAH in soil from 80 % at the beginning of the experiment to 39 % at its end and to the increase of the proportion of 5,6-nuclear polyarenes.

Table 5. Content of "extra-PAH" in soil, nM/g

PAH	Mass fraction of benz[a]pyrene added in soil, ng/g			
	10	20	30	40
∑PAH	0.036	0.098	0.070	0.148
∑ light PAH	0.025	0.047	0.008	0.021
Chrysene	0.025	0.036	0	0
∑ heavy PAH	0.011	0.051	0.062	0.127
Benz[b]fluoranthene	0	0.038	0.035	0.076
Benz[k]fluoranthene	0	0	0.002	0.012
Benz[ghi]perylene	0.011	0.013	0.022	0.014
Decrease of benz[a]pyrene	0.037	0.075	0.109	0.134

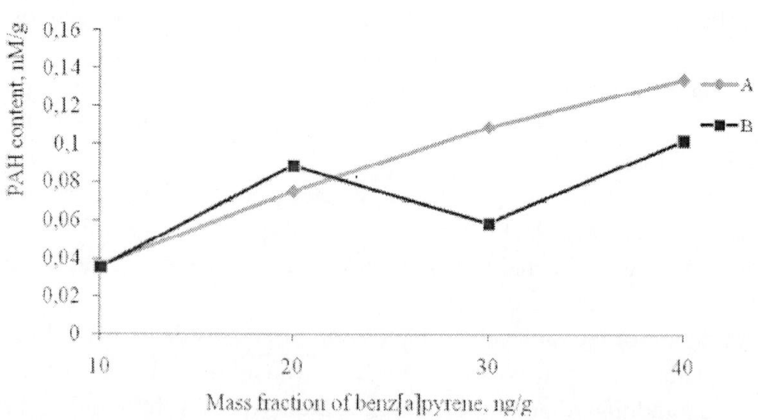

Figure 13. Contents of PAH at different benz[a]pyrene doses in soil. A – decrease of benz[a]pyrene, B – increase of benz[b]fluoranthene, benz[k]fluoranthene and benz[ghi]perylene.

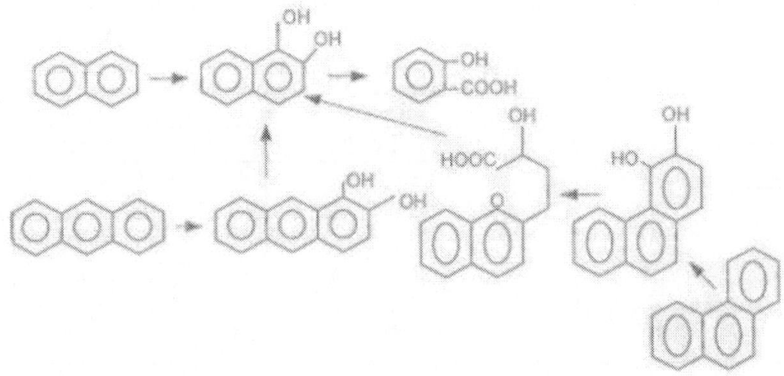

Figure 14. PAH oxidation as per E.S. Yelin, 2001.

The results of the examination of microbiological reactivity of soils polluted with benzpyrene allow to explain to some extent the appearance of "extra-PAH" in soil with different levels of pollution (Yakovleva et al., 2010). The injection of benzpyrene into soil caused the intensification of processes of soil organic matter mineralization, redistribution of micromicete populations according to the dominance degree, which, in its turn, resulted in the formation of various polyarenes in the process of microbiological degradation. The specific character of microbiological processes in soils was first of all determined by the weight fraction of benzpyrene injected into the soil. Under the influence of microbial enzymes complex processes of the transformation of PAH quantitative relation. Benzpyrene was a peculiar kind of catalyst of intersoil mineralization processes. The decomposition of PAH by microorganisms, as a rule, starts with hydroxylation of only one aromatic ring. With low doses of benzpyrene in soil (10 ng/g) microorganisms were able to decompose organic matters to light polyarenes, the increase of which we observed in soil (chrysene). With high benzpyrene doses soil accumulated polynuclear compounds, including heavy PAH. This process might be explained by the phenomenon of condensation of partially oxidated hydrocarbon. Benzpyrene oxidized in the least extent, than with lower doses. Micromicetes can also participate in the process of destruction of organic matter. They synthesize and emit into the environment various hydrolytic ferments breaking up any organic substrates. Fungi, just like microorganisms, decompose polyarenes by means of gradual hydroxylation of aromatic rings (Figure 14).

Microbial complex allows recording the effect of small doses of benzpyrene (10 ng/g) on the biological level. Microbial community reacted on small doses of benzpyrene by sharp decrease of their population. Micromicetes-indicators (species of genus *Paecilomyces* and *Fusarium moniliforme)* allowed detecting high doses (30-40 ng/g) of benzpyrene in soil.

PAH Content in Tradescantia (Clone 02)

In the course of chemical analysis in plants the following PAH were identified: phenanthrene, fluoranthene, pyrene, chrysene, benz [a] anthracene, benz [k] fluoranthene, benzpyrene and benz [ghi] perylene, benzO [b] fluoranthene, dibenz [a, h] anthracene, indenopyrene (table 5). The combination of polyarenes in plants was mostly represented by phenanthrene, fluoranthene, pyrene and chrysene, their weight fraction constituted 76-96 %.

Table 5. Mass fraction of polyarenes in whole plant, ng/g

PAH	Mass fraction of benz[a]pyrene added in soil, ng/g									
	0		10		20		30		40	
	$\overline{X}$*	$S\overline{x}$**	$\overline{X}$	$S\overline{x}$	$\overline{X}$	$S\overline{x}$	$\overline{X}$	$S\overline{x}$	$\overline{X}$	$S\overline{x}$
∑PAH	304.7	53.8	330.2	81.1	543.4	147	410.1	64.5	271.1	45.1
Phenanthrene	128.4	28.9	150.5	37.2	222.1	59.1	169.2	20.1	110.6	26.0
Fluoranthene	38.2	4	34.7	10	79.0	19.9	46.9	5	20.8	4.6
Pyrene	50.4	10	40.8	8.8	90.3	18.4	50.4	7.3	38.9	7.5
Chrysene	27.7	5.7	21.2	5	48.6	17.1	23.2	3.2	26.7	4.2
Benz[b]fluoranthene	0	0	0.3	0.3	55.4	9.7	13.0	1.3	28.9	18.0
Benz[k]fluoranthene	2.2	0.4	3.5	1.7	2.1	0.8	3.5	1	8.9	0.6
Benz[a]pyrene	2.2	0.17	6.0	0.3	8.1	0.51	11.8	1.5	15.8	2.4
Benz[ghi]perylene	9.8	3.1	13.4	3.8	13.5	0.8	17.2	3.4	20.1	2.1
∑ light PAH	290.5	52.3	307	73.9	464.3	88.6	364.6	53.9	214.6	36
∑ heavy PAH	14.2	3.2	23.2	4.4	79.1	69.7	45.5	14.2	56.5	16.7
Benz[a]pyrene, %	0.7	-	1.8	-	1.5	-	2.9	-	5.8	-
Fraction of 3, 4-nuclear PAH, % of ∑ PAH	95.4	-	92.9	-	85.4	-	88.9	-	79.2	-

The increase of weight fraction of benzpyrene in soil lead to the growth of the content of heavy hydrocarbons in plants. All polyarenes detected in plants can be divided into two groups according to their behaviour (Figure 15).

The first group – PAH that are represented in soil by their insignificant quantity and that are generated in significant quantities in plants (phenanthrene). The second group comprises polyarenes which the more were absorbed by a plant, the higher was the level of pollution induced by them. Polyarenes of the second group can be divided into PAH the content of which in plants increased with a particular dose (20 ng/g of benzpyrene in soil), after which the decrease in their accumulation could be observed (fluoranthene, pyrene, chrysene, benz[b]fluoranthene).

Polyarenes (benz[k]fluoranthene, benzpyrene, benz[ghi]perylene), the content of which in soil and plants grew with the increase of doses of injected benzpyrene. The effect of synthesis of phenanthrene by a plant with the accumulation of it identical to the change of tradescantia biomass was observed. In the similar way fluoranthene, pyrene, chrysene and benzO [b] fluoranthene were accumulated. It is known that the PAH synthesis by plants is maximal at the stage of dynamic vegetation, and after the period of intensive growth the synthesis of PAH in plants goes down (Frolovskaya et al., 1982).

As a result the summary content of polyarenes in plants increased within the interval of the benzpyrene injected from 0 to 20 ng/g of soil (Figure 16). The increase of benzpyrene doses from 30 to 40 ng/g of soil caused sharp decrease of polyarenes content in the whole plant, leaves, roots, as compared to deses of 10 - 20 ng/g of soil. The growth of polyarenes content in a plant depended on its biomass (Figure 5). In plants organs polycyclic aromatic hydrocarbons were distributed unevenly (Figure 17). The distribution was acropetal (polyarenes content increased from roots to leaves). The total content of polyarenes in leaves exceeded 1.5-3.1 times their content in roots. Maximal excess was observed with the pollution of 10 ng/g and in the reference sample.

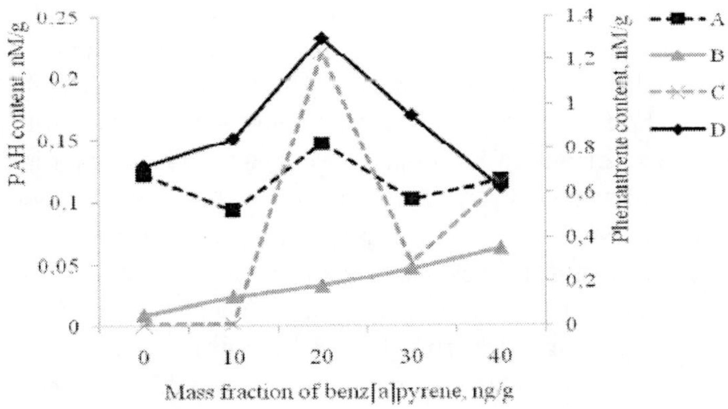

Figure 15. Content of PAH in plant: A – chrysene, B – benz[a]pyrene, C – benz[b]fluoranthene, D – phenanthrene.

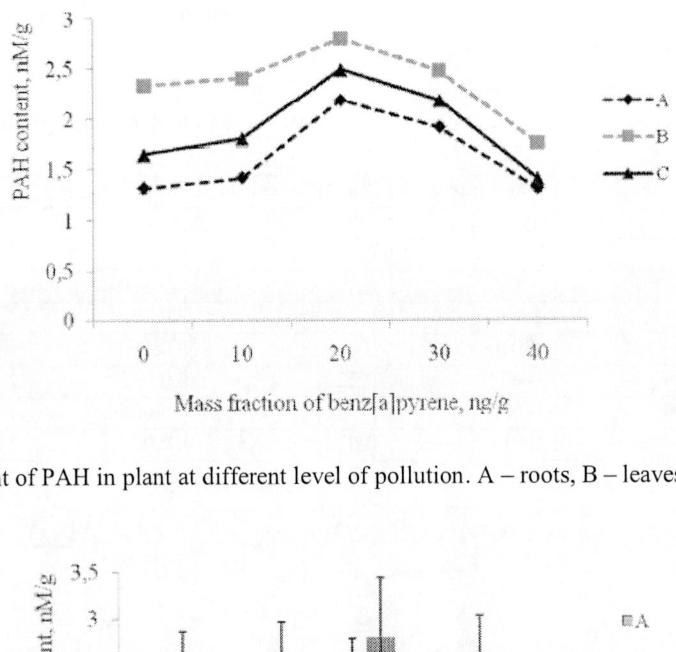

Figure 16. Content of PAH in plant at different level of pollution. A – roots, B – leaves, C – whole plant.

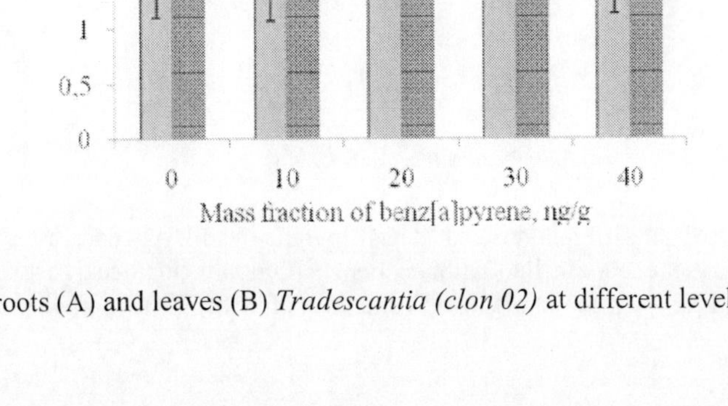

Figure 17. PAH in roots (A) and leaves (B) *Tradescantia (clon 02)* at different level of benz[a]pyrene soil pollution.

With the dose of 20 ng/g the differences were not reliable and were defined by a high content of phenanthrene in leaves, rather than in roots. A high content of polyarenes in plants was possibly connected with their biofriendliness and special character of their functional purpose for plants. PAH, as it seems, are essential components of plants which participate in metabolic processes. The amount of PAH detected in plants with the injection of benzpyrene in soil (table 6) depended on its dose and constituted 0.16 – 0.85 nM/g. It should be noted that with the pollution of soil with 10 ng/g of benzpyrene a significant increase of PAH in plants was observed as compared to the nil-treatment sample, with 20 ng/g – it was maximal. In both cases the increase of light PAH was determined by the growth of content of phenanthrene in plants, and not by the absorption from soil. The increase of heavy PAH was determined by absorption from soil. With the increase of benzpyrene dose to 30 – 40 ng/g the amount of additional PAH detected in plants decreased, although the existing increment was explained by the increase of the content of heavy PAH in comparison with the nil-treatment sample. It must be stated that the plant in general absorbed minimal amount of polyarenes from soil. The content of absorbed heavy PAH in plants was 10 times lower, than their content in soil. At that, the total increment of of heavy PAH in soil and in plants turned out to be a bit higher than the decrease of benzpyrene in soil (Figure 18).

Table 7. Contents of PAH founded in plant under benz[a]pyrene influence, nM/g

PAH	Mass fraction of benz[a]pyrene added in soil, ng/g			
	10	20	30	40
∑ PAH	0.160	0.850	0.520	-0.250
∑ light PAH	0.120	0.590	0.410	-0.480
Phenanthrene	0.120	0.580	0.230	-0.100
∑ heavy PAH	0.03	0.260	0.110	0.230
Benz[b]fluoranthene	0.002	0.220	0.038	0.110
Benz[k]fluoranthene	0.005	0	0.005	0.027
Benz[ghi]perylene	0.013	0.014	0.027	0.037

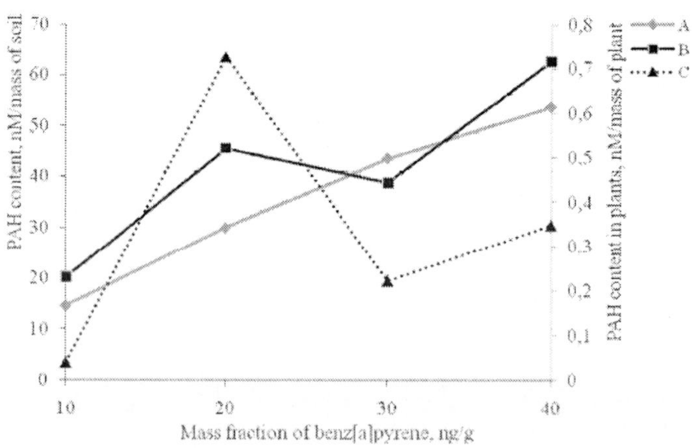

Figure 18. Content of PAH at different doses of benz[a]pyrene in soil. A – decrease of benz[a]pyrene in soil, B – increase of chrysene, benz[b]fluoranthene, benz[k]fluoranthene, benz[ghi]perylene in soil and heavy PAH including benz[a]pyrene in plants, C – increase of heavy PAH including benz[a]pyrene in plants.

It allows to say that the originated PAHs are the result of the decomposition of not only benzpyrene, but also other organic matter from soil. The content of chrysene in a plant depended little on the dose of the injected benzpyrene and remained on the level with the nil-treatment sample or less than that. The increase of its content can only be observed with the dose of 20 ng/g of benzpyrene, which corresponds to the maximal content of chrysene in soil and maximal biomass of plants.

So, the accumulation of polycyclic aromatic hydrocarbons in soils and plants occurred due to the application of benzpyrene into soil, transformation of the soil organic matter, and their intercellular synthesis in plants.

Polycyclic aromatic hydrocarbons are characterized by high similarity to humic and fulvic acids and as a result of donor-acceptor interrelations (Yashchenko et al., 1999; Kollist-Sigur K. et. al.,2001) they can serve as structural units for the formation of humus substances (Figure 19).

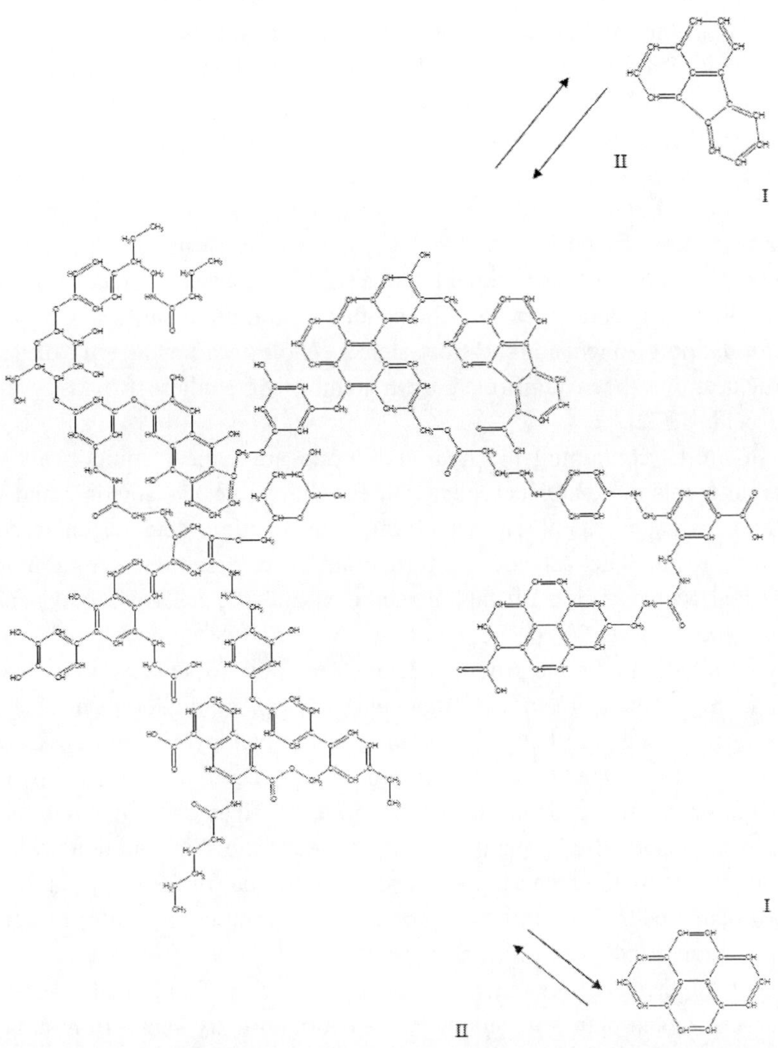

Figure 19. Scheme of fragmentary updating of humus molecules by polyaromatic hydrocarbons: I – fragment of humus molecule (as per D.S. Orlov), II – fluoranthene, III – pyrene.

The formation of molecules of polycyclic aromatic hydrocarbons in soilscan also occur due to transformation of cyclic isoprenoids, full or partial loss of side groups, dehydration, and activation by benzpyrene (catalyst) of mineralization of soil organic matter. Moreover, in the process of growth and development of plants roots and soil biota partially die away with the formation of various organic compounds, including pentacyclic terpanes, carbocyclic aromatic structures, as well as structures with diene and polyene bonds in aliphatic hydrocarbon chains, lipids, which can become sources of polyarenes.

Today it has been established that polycyclic aromatic hydrocarbons are included into the composition of soil lipids (Orlov, 1979; Lodygin et. al., 2005), in which phenanthrene, fluoranthene, pyrene, chrysene benzanthracene and many others were detected. The application of benzpyrene, as it seems, catalyzes all aforementioned processes in soils.

With the application of small amount of benzpyrene into soil first of all low-molecular polyarenes are formed. If the dose of benzpyrene applied into soil is increased to 30-40 ng/g, aas a result of transformation mostly high-molecular PAH appeared.

The intensity of bioaccumulation of polyarenes in plants was explained by the content of benzpyrene in soil. The consumption of polyarenes by a plant depended on the quantity of benzpyrene applied into soil within the range of 0 – 20 ng/g. The increase of benzpyrene doses from 30 to 40 ng/g of soil lead to sharp reduction of polyarenes in a plant owing to light polyarenes. For higher plants along with the absorption of polycyclic aromatic hydrocarbons from soil their intercellular synthesis in plants themselves is possible. The proof of that was the presence in plants of fluorine, naphthalene, anthracene, benz[a]anthracene which were absent in soil during the whole experiment, and also the increased content of phenanthrene in plants as compared to its content in soil. Phenanthrene in high quantities was synthesized by plants. The formation of phenanthrene from sterols (cholesterol) as a result of destruction of aliphatic structures has been confirmed by a number of studies (Rovinsky et al., 1988; Nesmeyanov et al., 1970).

Steroid structures containing the hydroxyl group are typical components of lipids of higher plants. In plants there are such phenanthrene derivatives, as abietinic and levopimaric acids. It is known that phenanthrene can form from abietinic acid which is contained by coniferous trees resin. The scheme of formation of phenanthrene from abietinic acid, including gradual aromatization of naphthenic rings and the loss of methyl substitutes, is given at Figure 20.

Chrysene in plants can be the product of transformation of pentacyclic triterpenoids with a hexatomic E-ring, which are main components of wax of higher plants. Therefore, the content of chrysene changed little with different doses of benzpyrene. Intensive accumulation of chrysene in a plant with the benzpyrene dose of 20 ng/g could be explained by its increased content in soil, or by the growth of plants biomass. Heavy PAH, excluding benz [b] fluoranthene, were absorbed by a plant from soil more, if the soil pollution with benzpyrene was higher. Dibenz [a, h] anthracene was almost not absorbed from soil by plants and was not synthesized in plants cells. The possibility for complex organic molecules to get into plants via the root system and their further assimilation in plants is a well-grounded fact corresponding to the theory of organic nutrition of plants (Zaalishvili et. al., 2000). Polycyclic aromatic hydrocarbons in plants are subject to transformation by means of hydroxylation and further buildup of dihydroxyacids derivatives and dicarboxylic acids. During the first stage of reductive-oxidative or hydrolytic transformation the molecule gets enriched with polar functional groups, which make it reactive and more soluble in water (Figure 14).

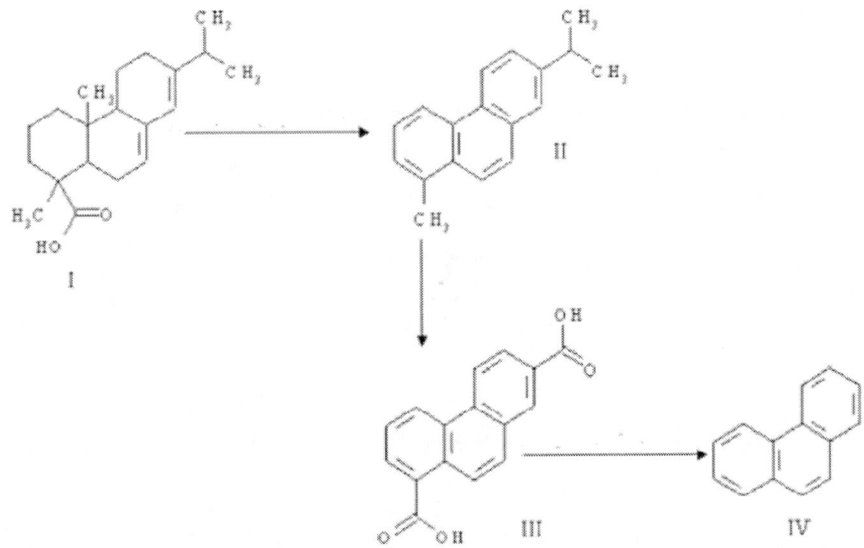

Figure 20. Theoretical mechanism of phenanthrene formation from coniferous trees resin components: I – abietinic acid, II – 1-methyl-7-isopropylphenanthrene, III – phenanthrene-1,7-dicarboxylic acid, IV – phenanthrene.

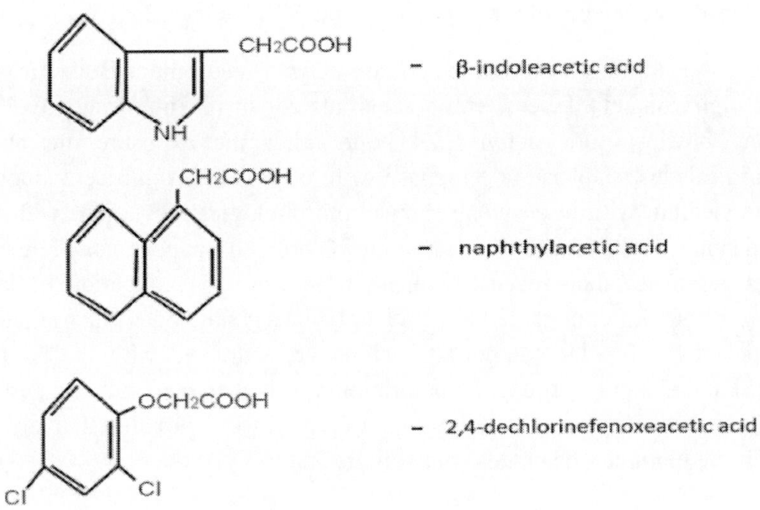

Figure 21. Auxin and its analogs, as per M.G. Safarov, 2001.

Carboxylated benzpyrene was able to have auxin effect. It was proved that under the influence of such herbicide as 2, 4 – dichlorophenoxyacetic acid the content of auxin in corn germs increase. In 1935 auxin activity was detected in naphthyl and naphtoilacetic acids (Figure 21). Benzpyrene can alco have such an effect. To possess auxin effect the compound must have an aromatic structure in the molecule content, to be an acid or to transform easily in acids in plants, to contain a methyl group between carboxyl and aromatic fragment (Baskakov et al., 1982, Mashtakov et al., 1967, Muromtsev et al., 1987). Metabolized benzpyrene is pretty likely to have such properties.

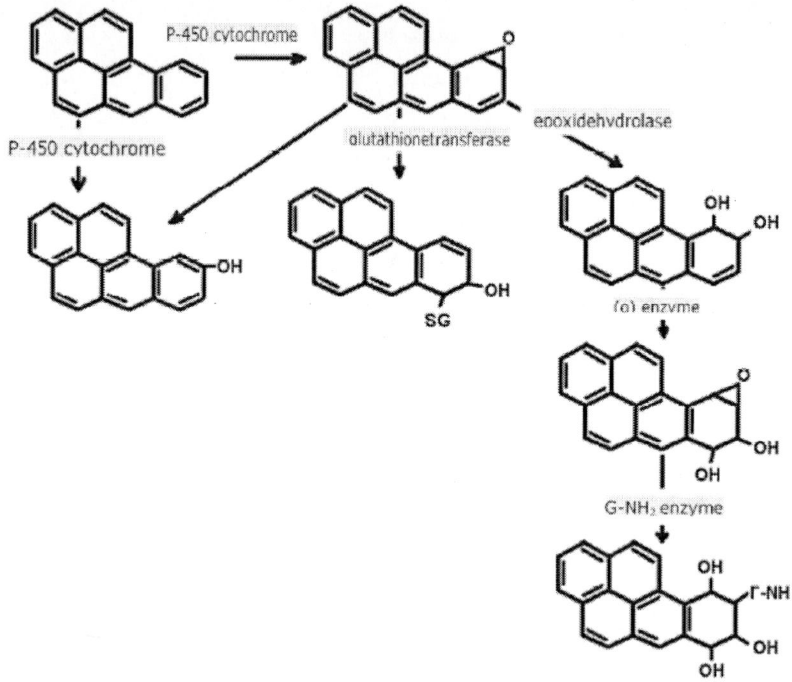

Figure 22. Scheme of a metabolism of benz[a]pyrene, as per V. I. Slesarev, 2001.

Krestovskaya et all. (1979) examined *Allium cera* – seed onion. Bulbs growing in the medium with high content of benzpyrene accumulated its maximal quantity. Benzpyrene started slowing down mitotic activity 24 hours after the exposure, no aberration of chromosomes in mitosis anaphase or telophase were observed. Insignificant amount of roots growing in the medium with benzpyrene showed morphological deviations within 10 hours: thickening and bends on the ends of roots, hypertrophy of cells, appearance of new side roots.

In the second phase there occur synthetic processes of conjugation of intermediary metabolism products with endogenic molecules: glycine, glutathione, glucose, etc., bringing about the appearance of polar compounds, which are extracted with the help of special mechanisms (Skugoreva et al., 2008). Transformation of benzpyrene with the participation of cytochrome P-450 can lead to the formation of epoxides. Epoxides actively join DNA fragments resulting in numerous pathologies (Figure 22).

Assessment of Toxic Activity of Benz[A]Pyrene in the Soil–Plant System

On the basis of the results obtained the total toxicological reactivity, which is the sensitive eco characteristic for the analysis of the degree of environmental pollution, of polycyclic aromatic hydrocarbons was calculated. The total toxicological reactivity was calculated according to the following formula:

$$T\sum_{i=1}^{n} PAH = \sum KTPAH_i \times \omega(PAH)_i,$$

where $\sum KTPAH_i$ – coefficient of toxicity of i PAH in reference to benzpyrene, (Nisbet et al., 1992), ω(PAH)i – weight fraction of i PAH in the object, ng/g.

Coefficients of PAH toxicity in reference to benz [A] pyrene: phenanthrene (0.001), anthracene (0.01), fluoranthene (0.001), pyrene (0.001), chrysene (0.01), benz [b]-fluoranthene (0.1), benz[k]fluoranthene (0.1), benzpyrene (1), dibenz [a,h] anthracene (1), benz [g,h,i] perilene (0.01). For the calculation of coefficients of toxicity median lethal dose (LD 50) characterizing major toxicity is used. It was established that toxicological reactivity of polyarenes for *Tradescantia (clone 02)* was explained by high-molecular PAH (Figure 23). The input of 5,6-nuclear polyarenes into toxicological reactivity of polycyclic aromatic hydrocarbons constituted 82÷99 %, benzpyrene had 40÷86 %, with its weight fraction in ∑PAH not exceeding 5 % for plants and 10 % for soil. The second after benzpyrene place in T∑PAH were occupied by benz[b]fluoranthene, benz[k]fluoranthene, benz[g,h,i]perilene.

For *Tradescantia* (clone 02) higher values of total toxicological reactivity of PAH were detected than for soil on which it was grown. The total toxicological reactivity of polycyclic aromatic hydrocarbons (Figure 24) in plants and in soil increased with the growth of pollution of soil with benzpyrene from 0 to 40 ng/g of soil.

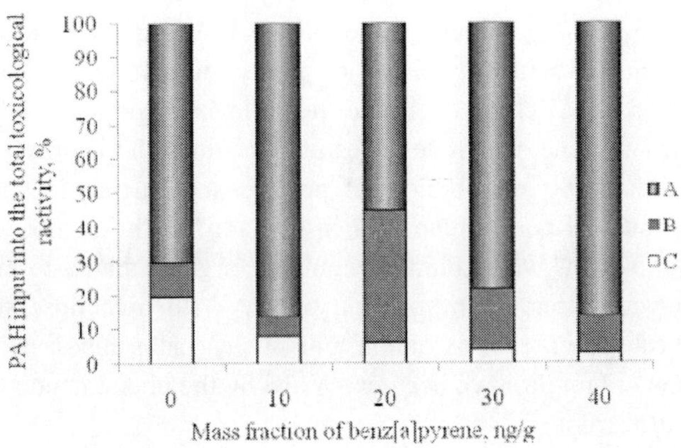

Figure 23. Contribution of heavy and light PAH to the toxicological PAH activity of plants at different level of benz[a]pyrene soil pollution. A – benz[a]pyrene, B – heavy PAH, C – light PAH.

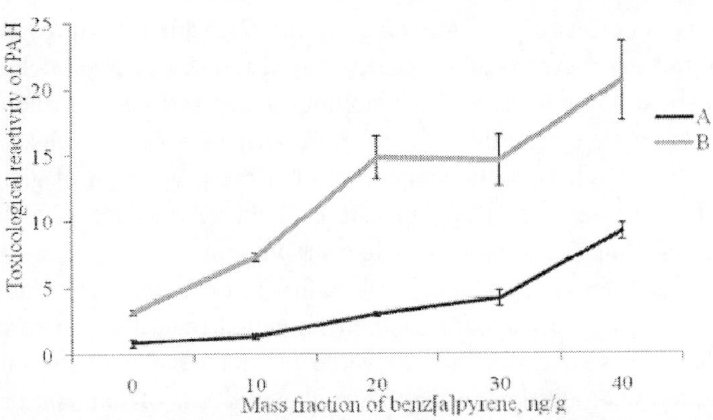

Figure 24. Toxicological activity of soil and plant PAH. A – soil, B – plants.

The excess of T∑PAH in polluted samples was 10 times higher in soil and 6 times higher in plants in comparison with the nil-treatment sample. In such a way, our studies showed that toxicological reactivity of PAH in soils and plants in the model experiment is determined mostly by heavy hydrocarbons and most of all by benzpyrene. This explained the choice of benzpyrene as the indicator of identified polycyclic aromatic hydrocarbons and as a marker of intensity of technogenesis impact on soils and plants. The data obtained allow suggesting that the reaction of plants on the cell level is defined owing to benzpyrene being the stress factor.

Conclusion

Growing of plants on soil polluted with benzpyrene causes the formation of a wide spectrum of polyarenes in soil. There was detected an effect of selective formation of PAH in soils explained by benz[a]pyrene dosage. The application of 10 – 20 ng/g of benz[a]pyrene into soil initiates the formation of mostly 4-nuclear polyarene – chrysene with the molar fraction in the total amount of light polyarenes of 95 %. With benz[a]pyrene doses of 30 – 40 ng/g additional formation of, to a greater extent, heavy 5,6-nuclear PAH occurs, among which the proportion of benz[b]fluoranthene makes 60 %. Additional selective formation of polyarenes can be connected with the activity of plants and soil microorganisms. Polyarenes are absorbed by Tradescantia (clone 02) more, if the level of pollution of soil with them is higher, excluding phenanthrene, which develops in plants in high quantity.

The specific character of microbiological processes occurring in soil was first of all determined by the amount of benzpyrene applied into soil. With low doses of benzpyrene in soil (10 ng/g) microorganisms were able to decompose organic matter to light polyarenes the increment of which was observed in soil (chrysene). With high doses of benzpyrene, in accordance with the rule of gradual oxidation, polynuclear substances including heavy PAH accumulated. This process might have been explained by the phenomenon of condensation of partially oxidized hydrocarbons.

PAHs are absorbed by plants and can cause various biological effects. Metabolites of benzpyrene, on the one hand, can serve as functional analogs of auxin and influence the growth of plants, and on the other hand, they cause various morphological, physiological and genetic transformations in plants. If benzpyrene is allied into soil in low doses its metabolism products produce auxin effect on *Tradescantia* (clone 02), which is expressed in the growth of roots biomass and the increase of frequency of appearance of giant cells in plants. The increase of benzpyrene doses leads to the inhibition of these processes. The growth of roots biomass was also observed during the cultivation of barley (Hordeum vulgare L.) on the soil polluted with BP. When soil is contaminated with oil, the proportion of giant cells is more than after the contamination with BP, which is evidently connected with the content of a significant amount of compounds with auxin-like structure in oil.

It was shown that the pollution of soil with oil or benzpyrene causes mutagenic, teratogenic and genotoxic effect in *Tradescantia* (clone 02), and this is expressed in proved increase of the level of somatic mutations, white mutant developments, inhibition of cell fragmentation and morphological abnormalities in the hair cells of staminal threads of plants. Non-linear dependence of the frequency of display of deviations in hairs of staminal threads of *Tradescantia* (clone 02) on the degree of soil pollution with oil was detected. Various levels of adaptation of *Tradescantia* (clone 02) to benzpyrene impact were found. On the morphological level the adaptation of plants was not explicit, adaptation was observed regarding physiological effects – white mutant developments and inhibition of cell fragmentation. On the level of genetic transformation in a plant new stages of adaptation are appearing, which corresponds to Selye's stress theory. The adaptation mechanisms of *Tradescantia* (clone 02) to oil pollution of soil to some extent start being reflected on the morphological level. The effects on physiological and genetic level are identical to effects initiated by benzpyrene.

It was shown that the main determinant of toxicological activity of deposit media (soils, plants) is the total content of 5,6-nuclear polyarenes, the input of which into toxicological activity makes 82-99 %.

ACKNOWLEDGMENTS

This work was partly supported by the Russian Foundation for Basic Research, project no. 11-04-00086-a and by the Ural Branch of the Russian Academy of Sciences, initiative project no. 12-U-4-1003.

REFERENCES

Barenboym, G. M., Ertsev, G. N. (2000). Some oil properties and oil effect on the environment and biota. Edited by V. S. Bibikova, A. P. Borovskikh. *The experience of oil spill liquidation in Usinsk district of the Republic of Komi*. P. 24-37.Syktyvkar: GUP Komimeleovodkhozproekt.

Baskakov, Y. U. A., Shapovalov, A. A. (1982). *Regulators of plants growth*. M.: Znaniye.

Becker, L., Glavin, D. P., Bada, J. L. (1997). Polycyclic aromatic hydrocarbons (PAHs) in Antarctic Martian meteorites, carbonaceous chondrites, and polar ice. *Geochimica et Cosmochimica Acta*. 1997. Vol. 61. N. 2. P. 475-481.

Belykh, L. I. (2005). Polycyclic aromatic hydrocarbons in natural and technogenic environment of the southern Baikal region. Geoecology. Geological engineering. *Hydrogeology. Geocryology*. №6. P. 539-551.

Belykh, L. I., Gorshkov, A. G., Ryabchikova, A. I., Seryshev, V. A., Marinayte, I. I. (2004). Distribution and biological reactivity of polycyclic aromatic hydrocarbons in the system «source – snow cover – soil – plant». *Contemporary problems of ecology*. № 6. P. 793-802

Bispo, A., Jourdain, M. J., Jauzein, M. (1999).Toxicity and genotoxicity of industrial soils polluted by polycyclic aromatic hydrocarbons (PAHs). *Organic Geochemistry*. Vol. 30. N. 8. P. 947-952.

Boonchan, S., Britz, M. L., Grant, A. S. (2000). Degradation and Mineralization of High-Molecular-Weight Polycyclic Aromatic Hydrocarbons by Defined Fungal-Bacterial Cocultures. *Applied and Environmental Microbiology*. Vol. 66, N. 3. P. 1007-1019.

Chernichenko, I. A., Pavlova, N. A. (1979). Some data on the transition of benz(a)pyrene-^{3}N from soil to plants. Edited by E. I. Slepyana. *Plants and chemical cancerogens.* P. 89-90. L.: Science.

Danetskaya, O. L. (1971). Experimental grounds of decancerization of cancerogenic industrial products. *L.: Medicine.* 221 p.

Danilov, A. M. (1996). Additives. The improvement of ecological characteristics of oil fuels. *M.: Chemistry.*

Dennis, J. A. (1976). Somatic aberration induction in Tradescantia occidentals by neutrons, X- and gamma-radiations. II Biological results, r. b. e. and o. e. r. *Int. J. Radiat. Biol. Relat. Stud. Phys. Chem. Med.* Vol. 29. N. 4. P. 323-342.

Dikun, P. P. (1979). Determination of polycyclic aromatic hydrocarbons. *Problems of analytical chemistry.* № 6. P. 100-116.

Dikun, P. P., Kalinina, I. A. (1979). Baseline content of benz(a)pyrene in a grain. Edited by E. I. Slepyana. *Plants and chemical cancerogens.* P. 113-115. L.: Science.

Dikun, P. P., Kostenko, L. D., Liverovsky, A. A., Shmulevskaya, E. I., Romanovskaya, L. S., Pankina, E. I. (1979). On mechanisms of formation of benz(a)pyrene during the wood pyrolysis. Edited by E. I. Slepyana. *Plants and chemical cancerogens.* P. 171-173. L.: Science.

Durmishidze, S. V., Buadze, O. A., Devdariani, T. V., Kakhaya, M. D., Kavtaradze, L. K. (1979). The influence of benz(a)pyrene on the ultrastructure of fragmented cells of the corn root. Edited by E. I. Slepyana. *Plants and chemical cancerogens.* P. 12-13. L.: *Science*, 1979.

Elin, E. S., Ignatova, V. A. (1997) Intermolecular relations of compounds containing phosphoryl group with phenols. *Journal of general chemistry. Issue* 7. №67. P. 1163-1165.

EPD F 16.1:2:2. 2:3. 39-03. (2003). Procedure of measurement of weight fraction of benz(a)pyreneA in samples of soil, ground, bed deposits and solid wastes by HELC method with the application of liquid chromatiograph «*Lyumachrom*». M. 27 p.

Eriksson, M., Sodersten, E., Zhongtang, Yu, Dalhammar, G., Mohn, W. W. (2003). Degradation of Polycyclic Aromatic Hydrocarbons at Low Temperature under Aerobic and Nitrate-Reducing Conditions in Enrichment Cultures from Northern Soils. *Applied and Environmental Microbiology.* Vol. 69. N. 1. P. 275-284.

Evseyeva, T. I., Geraskin, S. A. (2000). Cytogenetic effects of separate and joint effect of thorium-232 nitrate and potassium on tradescantia (clone 02). *Cytology.* Volume 42. №8. P. 822-828.

Evseyeva, T. I., Geraskin, S. A. (2001). The influence of combined factors of radiation and non-radiation nature on tradescantia. Ekaterinburg: Ural Branch of Russian Academy of Science, 2001.

Filchenkova, V. I. Biological activity of typical chernozem under the effect of oil pollution. *Preservation of plants and environmental protection in Tatarstan ASSR: collection of scientific works* P. 74-75. Kazan, 1982.

Filchenkova, V. I. Biological activity of typical chernozem under the effect of oil pollution. *Preservation of plants and environmental protection in Tatarstan ASSR: collection of scientific works* P. 74-75. Kazan, 1982.

Fismes, J., Perrin-Ganier, C., Empereur-Bissonnet, P., Morel, J. L. (2002). Soil-to-Root Transfer and Translocation of Polycyclic Aromatic Hydrocarbons by Vegetables Grown on Industrial Contaminated Soils. *Journal of Environmental Quality*. Vol. 31. P. 1649-1656.

Frolovskaya, V. N., Pikovsky, Yu. I., Teplitskaya, T. A., Alekseyeva, T. A., Ogloblina, A. I. (1982). Some aspects of geochemistry of polycyclic aromatic hydrocarbons. *Geochemistry of landscapes and soil geography*. P. 71-83. M.: Publishing House of the Moscow University.

Frolovskaya, V. N., Pikovsky, Yu. I., Teplitskaya, T. A., Alekseyeva, T. A., Ogloblina, A. I. (1982). Some aspects of geochemistry of polycyclic aromatic hydrocarbons. *Geochemistry of landscapes and soil geography*. P. 71-83. M.: Publishing House of the Moscow University.

Galiulin, R. R., Galiulin, R. V., Voznyak V. M. (2005). Distribution of benz(a)pyrene and arsenic in the environment. *Agrochemistry*. №5. P. 76-81.

Gao, Yan-zheng; Zhu, Li-zhong. (2005). Phitoremediation for phenanthrene and pyrene contaminated soils. *Journal of Environmental Science*. Vol. 17. N 1. P.14–18.

Gaponyuk, E. I. (1977). Remaining content of pesticides in environmental objects and their biological value. Pollution of atmosphere and soil. *Studies of IEM. Issue* 7(76). M.: Gidrometizdat. P. 65-88.

Gennadiev, A. N., Pikovsky, Y. U. I., Florovskaya, V. N., Alekseyeva, T. A., Kozin, I. S., Ogloblina, A. I., Ramenskaya, M. E., Teplitskaya, T. A., Shurubor, E. I. (1996). *Geochemistry of polycyclic aromatic hydrocarbons in rocks and soils*. M.: Publishing House of Moscow State University. 1996

Goncharov, I. V. (1987). *Geochemistry of West Siberia oils*. M.: Nedra.

Grant, W. F., Salamone, M. F. (1994). Comparative mutagenicity of chemicals selected for test in the International Program on Chemical Safetys collaborative study on plant systems for the detection of environmental mutagens. *Mutation Rescach*. Vol. 310. N 2. P. 156-162.

Guseynov, T. I., Alikperov, R. E. (1989). *Nature protection during sea oil fields exploration*. M.: Nedra.

HS.2.1.7.2041-06 (2006). Maximal permissible concentrations (MPC) of chemical substances in soil. Hygienic standards. M.: Federal center of hygiene and epidemiology of Rospotrebnadzor. 11 p.

Huang, X.-D., El-Alawi, Y., Penrose, D. M., Click, B. R., Greenberg, B. M. (2004) A multi-process phytoremediation system for removal of polycyclic aromatic hydrocarbons from contaminated soils. *Environmental Pollution*. Vol. 130. P. 465-476.

Ichikawa, S., Ishii, C. (1991). Validity of simplified scoring methods of somatic mutations in Tradescantia stamen hairs. *Environ. Exp. Bot.* vol. 31, N 2. P. 247 - 252.

Ichikawa, S., Sparrow, A. H. (1978). Influence of radiation exposure rate on somatic mutation frequency and loss of reproductive integrity in Tradescantia stamen hairs. *Mutat. Res.* Vol. 52. N 2. P. 195-211.

Jian, Y., Wang, L., Peter, P., F., Yu, H. T. (2004). Photomutagenicity of 16 polycyclic aromatic hydrocarbons from the US EPA priority pollutant list. *Mutat Res.* Vol. 557. P. 99– 108.

Kipopoulou, A. M., Manoli, E., Samara, C., (1999). Bioconcentration of polycyclic aromatic hydrocarbons in vegetables grown in an industrial area. *Environmental Pollution*. Vol.106. P. 369–380.

Kollist-Sigur, K., Nielsen, T., Gron, C., Hansen, P. E., Helweg, C., Jonassen, K. E. N., Jιιrgensen, O., Kirso, U. (2001). Sorption of Polycyclic Aromatic Compounds to Humic and Fulvic Acid HPLC Column Materials. *Journal of Environmental Quality*. Vol. 30. P. 526-537.

Krauss, M., Wilcke, W., Martius, C., Bandeira, A. G., Garcia, M. V. B., Amelung, W. (2005). Atmospheric versus biological sources of polycyclic aromatic hydrocarbons (PAHs) in a tropical rain forest environment. *Environmental Pollution*. Vol. 135. P. 143-154.

Krestovskaya, T. V., Slepyan, E. I., Kalinina, I. A. (1979). Influence of benz(a)pyrene on Allium Cera. Edited by E. I. Slepyana. *Plants and chemical cancerogens*. P. 38-41. L.: Science.

Kutsenko, S. A. (2002) *Principles of toxicology*. Saint-Petersburg.: Foliant, 2002.

Lodygin, E. D., Beznosikov, V. A. (2005). Influense of soil moisture on concentration and 13C NMR profiles of lipids in three Albeluvisoils. *Geoderma*. V. 127. N. 3-4. P. 253-262.

Ma, T. H., Kong, M. S. (1997). Genotoxicity of contaminated soil and well water detected by plant bioassays. *Environ. And Mol. Mutagenes*. Vol. 29. №28. P. 32.

Mashtakov, S. M., Deyeva, V. P., Volynets, A. P. (1967). Physiological effect of herbicides on cultivated plants sorts. Minsk: Science and technology.

Mc Kenna, J., Heath, R. D. (1976). Biodegradation of Polynuclear Aromatic Hydrocarbon Pollutants by Soil and Water Microorganisms. *Research Report*. N. 113. P.32.

Mericle, L. W., Mericle, R. P. (1967). Genetic nature of somatic mutations for flower color in Tradescantia, clone 02. *Rad. Bot*. N. 7. P. 449-464.

Merzlyak, M. N. (1989) Activated oxygen and oxidative processes in vegetative cell membranes. Results of science and technology. All-Russian Institute of Scientific andTechnical Information. *Series: Physiology of Plants. M: All-Russian Institute of Scientific and Technical Information*. Volume 6. P. 1-168.

Meudec, A., Dussauze, J., Deslandes, E., Poupart, N. (2006). Evidence for bioaccumulation of PAHs within internal shoot tissues by a halophytic plant arti.cially exposed to petroleum-polluted sediments. *Chemosphere*. Vol. 65. P. 474–481.

Minibaeva, F. V., Gordon, L. Kh. (2003) Products of superoxide and reactivity of extracellular peroxidase in vegetative tissues under stress. *Physiology of Plants*. № 3. P. 459-464.

Muromtsev, G. S., Chkanikov, D. I., Kulayeva, O. N., Gamburg, K. Z. (1987). Principles of chemical regulation of growth and productivity of plants. M.: Agropromizdat.

Nadala, M., Schuhmachera, M., Domingo, J. L. (2004). Levels of PAHs in soil and vegetation samples from Tarragona County, Spain. *Environmental Pollution*. Vol. 132. P. 1–11.

Nesmeyanov, A. N., Nesmeyanov, N. A. (1970). *Principles of organic chemistry*. M.: Chemistry.

Nielsen, T., Jorgensen, H. E., Larsen, J. C., Poulsen, M. (1996). City air pollution of polycyclic aromatic hydrocarbons and other mutagens: occurrence, sources and health effects. *The Science of the Total Environment*. Vol. 189-190. P. 41-49.

Orlov, D. S. (1979) Theoretical and practical problems of humus substance chemistry. Results of science and technology. All-Russian Institute of Scientific and Technical Information of Academy of Science of USSR. *Series: Soil Studies and Agrochemistry. M.: All-Russian Institute of Scientific and Technical Information*. Volume 2. P. 58-132.

Osipova, R. G., Shevchenko, V. A. (1984). Use of tradescantia (clone 02 and 4430) in the research of radiation and chemical mutagenesis. *Journal of general biology.* Volume 45.№2. P.226-332.

Panov, G. E., Petryashin, L. F., Lysyany, G. N. (1986). *Environmental protection at oil and gas enterprises.* M.: Nedra.

Pavlova, N. A., Donina, I. L. (1979) The meaning of benz(a)pyrene water solubility for its transition from soil into plants. Edited by E. I. Slepyana. *Plants and chemical cancerogens*. P. 99-100. L.: Science.

Polevoy, V. V. (1986). *The role of auxin in plants regulation systems*. L.: Science.

Polevoy, V. V., Salamatova, T. S. (1991). *Physiology of growth and development of plants: Study guide*. L.: Publishing house of the Leningrad university.

Raikhman, Ya. G., Nelyudin, V. A. (1999). Management of cancerogenic situation and prevention of cancer (comprehensive approach). Elista: Dzhangar.

Rovinsky, F. Ya., Teplitskaya, T. A., Alekseyeva, T. A. (1988). *Baseline monitoring of polycyclic aromatic hydrocarbons*. L.: Hydrometeoizdat.

Rubtsova, N. A., Troyanskaya, A. F., Moseyeva, D. P. (1997). Polycyclic aromatic hydrocarbons in bed deposits of the North Dvina and Dvina Bay. *Environmental Chemistry*. № 3. P. 151-157.

Safarov, M. G. Herbicides: 2,4-d. (2001). *Chemistry*. Volume 7. № 9. P. 57-62.

Selye, G. (1977). Growth concept. The news about hormones and mechanisms of their activity. Kiev: Naukova Dumka.

Shabad, L. M., Khesina, A. Ya. (1979). On the possibility of endogenic synthesis of benz(a)pyrene in plants seeds at seed germination. Edited by E. I. Slepyana. *Plants and chemical cancerogens*. P. 86. L.: Science.

Shabad, L. M., Khesina, A. Ya., Smirnov, G. A. (1979). Exhaust gases of cars as the main source of atmospheric pollution of cities with cancerogenic hydrocarbons. Edited by A. P. Ilyitsky, A. I. Shilina. *Cancerogenic substances in the environment*. P. 29-34. M.: Hydrometeoizdat.

Shurubor, E. I. (2000). Polycyclic aromatic hydrocarbons in the system soil-plant in the oil processing area (the Perm Kama region). *Soil studies*. № 12. P. 1509-1514.

Simonich, S. L., Hites, R. A. (1994). Imprortanse of vegetation in removing polycyclic aromatic hydrocarbons from the atmosphere. *Natura*. 1994. V. 370. N. 7. P.49-51.

Simonich, S. L., Hites, R. A. (1995). Organic pollutant accumulation in vegetation. *Environ. Sci. and Tecnology*. V.29. N. 12. P. 2905-2914.

Skugoreva, S. G., Ogorodnikova, S. Yu., Golovko, T. K., Ashikhmina, T. Ya. (2008). Phytotoxicity of phosphor organic compounds and mercury. Ekaterinburg: Ural Branch of Russian Academy of Science.

Slepyan, E. I., Sheynerman, N. A., Andreyeva, Z. A., Firsov, G. A., Erik, V. A., Kolchevsky, A. G. (1979). The content of benz(a)pyrene in falling leaves of *Crataegus curonica, Viburnum opulus, Salix aurita, Betula verrucosa* and in needles of *Pinus silvestris*. Edited by E. I. Slepyana. *Plants and chemical cancerogens*. P. 177-179. L.: Science. Slesarev, V. I. (2001) *Chemistry: principles of living chemistry: Course book for higher educational institutions*. -2nd edition, amended and enlarged. Saint-Petersburg: Khimizdat.

Smith, M. J., Flowers, T. H., Duncan, H. J., Alder, J. (2006). Effects of polycyclic aromatic hydrocarbons on germination and subsequent growth of grasses and legumes in freshly contaminated soil and soil with aged PAHs residues. *Environmental Pollution*. Vol. 141. P. 519 -525.

Sorokin, YA. T. (1983) *Waste-free production in oil processing industry*. M.: Chemistry.

Sparrow, A. K. H., Lloyd, A., Schairer, L. (1977). Appearance of somatic mutations in Tradescantia under the influence of chemical mutagens EMS (ethyl methane sulphonate) and DBE (1,2-dibromoethane) and specific atmospheric pollutants O_3, SO_2, NO_2. Edited by N. P. Dubinin. *Genetic consequences of environmental pollution*. P. 50-61. M.: Science.

Tao, S., Cui, Y. H., Xu, F. L., Li, B. G., Cao, J., Liu, W. X., Schmitt, G., Wang, X. J., Shen, W. R., Qing, B. P., Sun, R. (2004). Polycyclic aromatic hydrocarbons (PAHs) in agricultural soil and vegetables from Tianjin. *Science Total Environ*. Vol. 320. P.11–24.

Tonkopy, N. I., Rozanova, V. Ya., Mazel, Yu. Ya. (1979). Experimental studies of the possibility of benz(a)pyrene fixing by plants. Edited by E. I. Slepyana. *Plants and chemical cancerogens*. P. 97-99. L.: Science.

Tonkopy, N. I., Shestopalova, G. E., Rozanova, V. Ya. (1979). Some factors determining the degradation of benz(a)pyrene in soil. Edited by A. P. Ilyitsky, A. I. Shilina. *Cancerogenic substances in the environment*. P. 65-68. M.: Hydrometeoizdat.

Ugrekhelidze, D. Sh. (1976). *Metabolism of exogenous alkanes and aromatic hydrocarbons in plants*. Tbilisi: Metsniereba.

Underbrink, A. G., Schaires, L. A., Sparrow, A. H. (1973). Tradescantia Stamen Hairs: A radiobiological test system applicable to chemical mutagenesis. *Chemical mutagenesis: Principles and Methods for their detections*. Vol. 3. N. 4. P. 159-175.

Velde, I. A., Rakhu, M. A., Itra A. R., Ilnitsky, A. P., Paalme L. P. (1979). The content of benz(a)pyrene in soil and plants of Estonia. Edited by E. I. Slepyana. *Plants and chemical cancerogens*. P. 101. L.: Science.

Volikov, A. N. (1989). *Combustion of gas and liquid fuel in low capacity boilers*. L.: Nedra.

Xu, S. Y., Chen, Y. X., Wu, W. X., Wang, K. X., Lin, Q., Liang, X. Q. (2006). Enhanced dissipation of phenanthrene and pyrene in spiked soils by combined plants cultivation. *Science of the Total Environment*. Vol. 369. P. 206 - 215.

Yakovleva, E. V., Beznosikov, V. A., Kondratenok, B. M., Khomichenko, A. A. (2011) Genotoxic effects in *Tradescantia* plants (clon 02) induced by benz(a)pirene. *Contemporary problems of ecology*. N 6. P. 805–812.

Yakovleva, E. V., Gabov, D. N., Beznosikov, V. A., Kondratenok, B. M., Vasileich M. I. (2008). Bioaccumulation of polycyclic aromatic hydrocarbons in the system soil-plant. *Agrochemistry*. № 9. P. 66-74.

Yakovleva, E. V., Khabibulina, F. M., Vinogradova, Yu. A., Beznosikov, V. A., Kondratenok, B. M. (2010). Microbiological activity of soils, polluted with benz(a)pyrene. *Agrochemistry*. № 11, P. 63-69.

Yashchenko, N. Yu., Perminova, I. V., Petrosyan, V. S., Filippova, E. M., Fadeyev, V. V. (1999). Interrelation of humus acids of various origin with polyaromatic hydrocarbons: impact of RN and ionic medium of the environment. Bulletin of the Moscow University. Series: 2. *Chemistry*. Volume 40. № 3. P. 188-193.

Zaalishvili, G., Khatisashvili, G., Ugrekhelidze, D., Gordeziani, M., Kvesitadze, G. (2000). Plant potential for detoxification (Review). *Applied Biochemistry and Microbiology*. Vol. 36. P. 443-451.

Zaynullin, V. G. (1998). Genetic effects of chronic ionizing radiation in low doses. Saint-Petersburg.: Science.

Zheng, Z., Obbard, J. P. (2002). Polycyclic Aromatic Hydrocarbon Removal from Soil by Surfactant Solubilization and Phanerochaete chrysosporium Oxidation. *Journal of Environmental Quality*. Vol. 31. P. 1842-1847.

BIOGRAPHICAL SKETCH OF AUTHORS

Name: Evgenia V. Yakovleva

Affiliation: Institute of Biology of Komi Scientific Centre of the Ural Branch of the Russian Academy of Sciences

Date of Birth: 01/09/1983

Education: Higher education: Syktyvkar State University, department of chemistry and biology, graduate (2000-2005). Postgraduate education: Komi Scientific Centre of the Ural Branch of the Russian Academy of Sciences, soil department, candidate of science (2005-2008).

Address: 167982, Russia, the Komi Republic, Syktyvkar, Kommunisticheskaya st., 28

Research and Professional Experience: analytical work to determine PAHs in soils and plants, study of PAHs in soils and plant.

Professional Appointments: laboratory technician, analyst, researcher

Honors: -Publications Last 3 Years:1. BIOACCUMULATION OF POLYCYCLIC AROMATIC HYDROCARBONS IN THE SOIL–PLANT SYSTEMS OF THE NORTHERN TAIGA BIOCENOSES. E. V. Yakovleva, V. A. Beznosikov, B. M. Kondratenok, and D. N. Gabov. ISSN 1064 2293, Eurasian Soil Science, 2012, Vol. 45, No. 3, pp. 309–320.42. GENOTOXIC EFFECTS IN *TRADESCANTIA* PLANTS (CLON 02) INDUCED BY BENZ(A)PIRENE. E.V. Yakovleva, V.A. Beznosikov, B.M. Kondratenok, and A.A. Khomichenko. ISSN 0869-8619, Contemporary problems of ecology, 2011, No. 6, pp. 805–812.3. POLYCYCLIC AROMATIC HYDROCARBONS IN THE ATMOSPHERIC PRECIPITATION–SOIL–LISYMETRIC WATERS SYSTEM OF TECHNOGENIC LANDSCAPES Gabov D.N., Beznosikov V.A., Kondratenok B.M., Yakovleva E.V. Conference Proceedings Book of Abstracts. 6th ISMOM 2011. International Symposium of Interactions of Soil Minerals with Organic Components and Microorganisms Soil Interfaces in a Changing World. 26th June – 1st July 2011, Montpellier, France. P. 191.

4. POLYCYCLIC AROMATIC HYDROCARBONS IN THE SOILS OF TECHNOGENIC LANDSCAPES Gabov D.N., Beznosikov V.A., Kondratenok B.M., Yakovleva E.V. Geochemistry International. 2010. T. 48. № 6. C. 569-579.

Name: Dmitriy N. Gabov

Affiliation: Institute of Biology of Komi Scientific Centre of the Ural Branch of the Russian Academy of Sciences

Date of Birth: 20/05/1979

Education: Higher education: Syktyvkar State University, department of chemistry and biology, graduate (1996-2001). Postgraduate education: Komi Scientific Centre of the Ural

Branch of the Russian Academy of Sciences, soil department, candidate of science (2001-2004).

Address: 167982, Russia, the Komi Republic, Syktyvkar, Kommunisticheskaya st., 28

Research and Professional Experience: analytical work to determine PAHs in soils and surface waters, study of PAHs and alkanes in soils, atmospheric precipitations, surface and lisymetric waters and wastes.

Professional Appointments: laboratory technician, analyst, researcher

Honors: - Publications Last 3 Years:

1. POLYCYCLIC AROMATIC HYDROCARBONS IN THE SOILS OF TECHNOGENIC LANDSCAPES Gabov D.N., Beznosikov V.A., Kondratenok B.M., Yakovleva E.V. Geochemistry International. 2010. T. 48. № 6. C. 569-579.

2. SATURATED HYDROCARBONS IN THE BACKGROUND AND CONTAMINATED SOILS OF THE CISURALS Gabov D.N., Beznosikov V.A., Kondratenok B.M., Gruzdev I.V. Eurasian Soil Science. 2010. T. 43. № 10. C. 1102-1108.

3. POLYCYCLIC AROMATIC HYDROCARBONS IN THE ATMOSPHERIC PRECIPITATION–SOIL–LISYMETRIC WATERS SYSTEM OF TECHNOGENIC LANDSCAPES Gabov D.N., Beznosikov V.A., Kondratenok B.M., Yakovleva E.V. Conference Proceedings Book of Abstracts. 6th ISMOM 2011. International Symposium of Interactions of Soil Minerals with Organic Components and Microorganisms Soil Interfaces in a Changing World. 26th June – 1st July 2011, Montpellier, France. P. 191.

Name: Vasiliy A. Beznosikov

Affiliation: Institute of Biology of Komi Scientific Centre of the Ural Branch of the Russian Academy of Sciences

Date of Birth: 18/06/1947

Education: Higher education: Perm agricultural institute, Faculty of soil science and agro chemistry, graduate (1966-1971). Postgraduate education: Perm agricultural institute, Faculty of soil science and agro chemistry, doctor of science (1972-1975).

Address: 614990, Russia, Syktyvkar, Kommunisticheskaya st, 23.

Head of the laboratory of soil chemistry of the Institute of biology of Komi Scientific Centre of the Ural Branch of the Russian Academy of Sciences (1976-2012).

Research and Professional Experience: study of the humus acids used ^{13}C-NMR and EPR spectroscopy; content of n-alkans, PAHs in soils.

Professional Appointments: Manager of laboratory of soils chemistry

Publications Last 3 Years:

1. SATURATED HYDROCARBONS IN THE BACKGROUND AND CONTAMINATED SOILS OF THE CISURALS Gabov D.N., Beznosikov V.A., Kondratenok B.M., Gruzdev I.V. Eurasian Soil Science. 2010. T. 43. № 10. C. 1102-1108.

2. POLYCYCLIC AROMATIC HYDROCARBONS IN THE ATMOSPHERIC PRECIPITATION–SOIL–LISYMETRIC WATERS SYSTEM OF TECHNOGENIC LANDSCAPES Gabov D.N., Beznosikov V.A., Kondratenok B.M., Yakovleva E.V. Conference Proceedings Book of Abstracts. 6th ISMOM 2011. International Symposium of Interactions of Soil Minerals with Organic Components and Microorganisms Soil Interfaces in a Changing World. 26th June – 1st July 2011, Montpellier, France. P. 191.

3. THE MOLECULAR STRUCTURE AND ELEMENTAL COMPOSITION OF HUMIC SUBSTANCES FROM ALBELUVISOLS. Lodygin E.D., Beznosikov V.A. Chemistry and Ecology, 2010. T. 26. C. 87-95.

4. ECOLOGICAL GEOCHEMICAL ASSESSMENT OF HYDROCARBONS IN SOILS OF NORTHEASTERN EUROPEAN RUSSIA. Beznosikov V.A., Lodygin E.D. Eurasian Soil Science, 2010. Vol. 43. № 5. P. 550-555.

Name: Boris M. Kondratenok

Affiliation: Institute of Biology of Komi Scientific Centre of the Ural Branch of the Russian Academy of Sciences

Date of Birth: 16/11/1955

Education: Higher education: Leningrad State University, chemistry department, graduate (1973-1978). Postgraduate education: Leningrad State University, chemistry department, candidate of science (1978-1981).

Address: 167982, Russia, Komi Republic, Syktyvkar, Kommunisticheskaya st., 28

Research and Professional Experience: Analytical chemistry, physical and chemical methods of the analysis, chemical environmental problems.

Professional Appointments: Deputy director of Institute of Biology of Komi Scientific Centre, manager of Ecoanalit laboratory

Honors: - Publications Last 3 Years:

1. POLYCYCLIC AROMATIC HYDROCARBONS IN THE SOILS OF TECHNOGENIC LANDSCAPES Gabov D.N., Beznosikov V.A., Kondratenok B.M., Yakovleva E.V. Geochemistry International. 2010. T. 48. № 6. C. 569-579.

2. SATURATED HYDROCARBONS IN THE BACKGROUND AND CONTAMINATED SOILS OF THE CISURALS Gabov D.N., Beznosikov V.A., Kondratenok B.M., Gruzdev I.V. Eurasian Soil Science. 2010. T. 43. № 10. C. 1102-1108.

3. POLYCYCLIC AROMATIC HYDROCARBONS IN THE ATMOSPHERIC PRECIPITATION–SOIL–LISYMETRIC WATERS SYSTEM OF TECHNOGENIC LANDSCAPES Gabov D.N., Beznosikov V.A., Kondratenok B.M., Yakovleva E.V. Conference Proceedings Book of Abstracts. 6th ISMOM 2011. International Symposium of Interactions of Soil Minerals with Organic Components and Microorganisms Soil Interfaces in a Changing World. 26th June – 1st July 2011, Montpellier, France. P. 191.

Name: Maria V. Aniskina

Affiliation: Syktyvkar State University

Date of Birth: 23/04/1980

Education: Higher education: Syktyvkar State University, department of chemistry and biology, graduate (1998-2002). Postgraduate education: Komi Scientific Centre of the Ural Branch of the Russian Academy of Sciences, department of radioecology, candidate of science (2002-2005).

Address: 167982, Russia, Komi Republic, Syktyvkar, Petrozavodskaya St., 120, 207th office.

Research and Professional Experience: influences of oil pollution on vegetative systems;

Professional Appointments: teacher, researcher;

Publications Last 3 Years: -

In: Pyrene
Editors: Petr Ruzicka and Tomas Kral

ISBN: 978-1-62417-291-5

Chapter 2

BENZO[*A*]PYRENE AND COLONIC CANCER

Atsushi Hakura* and Jiro Sonoda
Drug Safety Japan, Global Drug Safety, Eisai Co., Ltd.
5-1-3 Tokodai, Tsukuba, Ibaraki, Japan

ABSTRACT

Colorectal cancer (CRC) is one of the most common cancers in the Western world. Subsets of food-derived heterocyclic amines and *N*-nitroso compounds are probably the environmental chemicals that cause CRCs in humans. However, the contribution of environmentally–ubiquitous polyaromatic hydrocarbons (PAHs) toward causing human CRCs is still unclear. One of the probable reasons for this is that not so much attention has been paid to the colonic carcinogenicity of PAHs. The number of PAHs that have been examined is extremely limited for colonic carcinogenicity induced by oral treatment (by gavage or in the diet). The most representative PAH that is carcinogenic is benzo[*a*]pyrene (BP). The carcinogenicity of BP has not been noted in the colon of rodents, although forestomach, spleen, lung, and liver are shown to be target organs for the carcinogenesis of BP by oral treatment. However, we showed that BP was mutagenic in multiple organs of mice, including the colon with the highest mutation frequency. Recently, we demonstrated that BP can rapidly induce colonic adenocarcinomas in mice at a high incidence/multiplicity in the presence of dextran sulfate sodium, which is non-genotoxic and induces colitis in the murine colon. Furthermore, we examined the mutagenicity of BP with the Ames test, a routine *in vitro* mutagenicity test used world–wide, using multiple human liver S9 fractions in addition to liver S9 fractions from rats pretreated with inducers of drug–metabolism enzymes. BP was weakly or marginally mutagenic in the majority of human S9 fractions; however, some of them showed potent mutagenicity with individually diverse mutagenic activities. These findings may partially account for the induction of colonic tumors by BP in humans, particularly to patients with inflammatory bowel disease. Thus, further studies of the carcinogenic risk of PAHs including BP on the colon are strongly recommended.

---

* To whom correspondence should be addressed; Tel: +81-298-47-6517; Fax: +81-298-47-6931; E-mail: a-hakura@hhc.eisai.co.jp.

INTRODUCTION

Colorectal Cancer (CRC) in Humans

CRC is one of the most common cancers in humans in the Western world [1, 2]. In the United States alone, 143,460 new cases of CRC have been diagnosed, and 51,690 deaths were attributed to this cancer [3]. Although the incidence and mortality are higher in countries with a Western lifestyle, the global incidence is rising, and the majority of the world's cases of CRC occur outside of countries in which traditional Western lifestyles are dominant [2]. Occurrence of CRC can be either hereditary (familial) or sporadic [1, 2, 4]. Of the familial forms of CRC, familial adenomatous polyposis and hereditary nonpolyposis colon cancer are the most common [1, 2, 4]. The former and latter are associated with germ line mutations in the tumor suppressor genes encoding proteins, such as adenomatous polyposis coli (APC), and DNA mismatch repair enzymes, respectively [1, 2, 4]. However, more than 95% of CRCs are considered to be sporadic and generated primarily by the accumulation of mutations in the multiple genes associated, which result from DNA damage, produced by environmental and endogenous factors [1, 2, 4-12, 14-16].

Diet as Probable Environmental Factors Causing or Modulating Human CRC

Epidemiological studies suggest that diet plays an important role in CRC susceptibility and is estimated to contribute to 80% of the CRC cases [4-12]. The remarkable increase in the incidence of CRC in Japan is well established to be associated with the westernization of lifestyle, especially in dietary habits [14-16]. Among Japanese in Japan, the increase in CRC incidence offers a sharp contrast to the decrease in gastric tumors, and this is similar among descendents of Japanese immigrants in the United States of America [14-16].

Dietary components causing or modulating human CRCs are suggested to include food-derived heterocyclic amines (HCAs) such as 2-amino-1-methyl-6-phenylimidazo[4,5-*b*]pyridine (PhIP) and 2-amino-3,4-dimethylimidazo[4,5-*f*]quinoline (MeIQ), *N*-nitroso compounds found in foods containing nitrite or beer, polyaromatic hydrocarbons (PAHs), fat, fecapentaenes, and bile acids [5-12]. Food-derived HCAs, *N*-nitroso compounds, PAHs, and fecapentaenes have a mutagenic potency to initiate CRCs [6, 8, 9, 11, 12], while fat and bile acids are considered to function primarily as promoters or modulators of colorectal carcinogenesis [7, 8, 10]. Adipose tissue might have another role of retaining hydrophobic chemicals or supplying peroxides in colorectal carcinogenesis [17, 18].

PAHs as Environmental Contaminants

PAHs including benzo[*a*]pyrene (BP) constitute a large chemical class of widespread environmental contaminants, and are formed during incomplete combustion or pyrolysis of organic materials [1,11,12,19]. They are present in air, water, soil, sediments, and foods [1,11,19]. Major sources of PAHs in ambient air include residential and commercial heating with wood, coal or other biomasses, other indoor sources such as cooking and tobacco smoke, and outdoor sources like motor-vehicle exhaust, industrial emissions and forest fires [1, 11, 19]. Average concentrations of individual PAHs in the ambient air in urban areas typically range from 1 to 30 ng/m^3 [19].

Estimates of PAH intake from food vary widely, ranging from a few nanograms to a few micrograms per person per day [19]. In Italy, dietary exposure to PAHs (3 μg/day) was estimated to be significantly higher than the respiratory intake of PAHs from polluted urban air (370 ng/day) [11]. Median dietary intake of 3 μg PAHs/day represents around 96% of the total daily exposure for nonsmokers; tobacco smoke adds around 2-5 μg/day for a one pack per day smoker, and up to 15 μg/day for a heavy smoker (3 packs per day) [11]. Sources of PAHs in the diet include barbecued/grilled/broiled and smoke-cured meats; roasted, baked and fried foods (high-temperature processing); bread, cereals and grains (at least in part from gas/flame-drying of grains); and vegetables grown in contaminated soils, or in areas with surface contamination from atmospheric PAH fall-out [11, 19]. Leaf vegetables can be a significant source of PAHs in the human diet [11].

Many PAHs are mutagens/clastogens in *in vitro* genotoxicity tests and are also rodent carcinogens, depending on their chemical structure [1, 11, 19, 21-26]. Such PAHs are considered to be associated with the induction of lung or skin cancers in humans through inhalation or *via* skin contact [1, 11, 19]. However, the contribution of PAHs to human CRC is still unclear. One of the probable reasons for this is that much less attention has been paid to the colonic carcinogenicity of PAHs as compared with other environmental chemicals such as food-borne HCAs and *N*-nitroso compounds [1, 6, 8].

BP as the Most Representative Carcinogenic PAH

BP is the most representative PAH, and is potently genotoxic and carcinogenic, as with many other PAHs [1, 4, 19, 22-26]. Recently, BP has been classified into Group 1 (human carcinogen) by the International Agency of Research on Cancer (IARC) [19]. Humans are exposed to BP, as with many other PAHs, through dietary intake (2-500 ng/day) and inhalation (10-50 ng/day) [1, 11]. It has been estimated that the life-long intake (*i.e.* 70 years) of BP for non-occupationally exposed humans may add up to 29 mg, integrating respiratory, gastrointestinal and percutaneous absorption [9]. Human breast milk is also reported to contain BP at 6.5 ng/kg levels [11]. Many fresh vegetables typically contain levels of 2.85-24.5 ppb BP [11].

van Maanen *et al.* [27] reported that the mean daily BP intake of 1.5 μg from hamburgers resulted in the formation of a predominant BPDE-dG adduct in white blood cells at levels varying between 3 and 103 adducts/10^{10} nucleotides on the 3rd day of the consumption period in 8 out of 21 subjects. Alexandrov *et al.* [28] detected BPDE-dG adducts in 4 out of 7 human surgical colon samples, with the adduct levels varying between 0.3 and 1.0 adducts/10^8 nucleotides, possibly derived from diet and/or tobacco smoke.

Current Understanding of Mechanisms Underlying BP-Induced Carcinogenesis

The current understanding of mechanisms underlying BP-induced carcinogenesis is summarized in Figure 1, and can be generally classified into: (1) reactive metabolite mechanism, (2) aryl hydrocarbon-receptor (AhR) mechanism, (3) epigenetic mechanisms, and (4) immunosuppression mechanism. The reactive metabolite mechanism is further divided into reactive fashions *via* "diolepoxide," "quinones," "radical cations," and "*meso*-region" [19]. Of these mechanisms, the diolepoxide theory is the most central and widely accepted. However, the contribution of each mechanism on BP-carcinogenesis may be variable, depending on the species and strains of animals, routes and amounts of exposure, organs/tissues, and other factors such as co-exposure with other chemicals.

(1) Reactive metabolite mechanism
- *via* diolepoxide (Bay-region theory)
- *via* quinones
- *via* radical cations
- *via meso*-region

(2) Aryl hydrocarbon-receptor (AhR) mechanism
(3) Epigenetic mechanisms
(4) Immunosuppression mechanism

Figure 1. Proposed mechanisms underlying BP-induced carcinogenesis.

Diolepoxide Mechanism (Bay-Region Theory)

The diolepoxide mechanism for BP features a sequence of metabolic transformation leading from BP to the ultimate activated form, *anti*-benzo[*a*]pyrene-*trans*-7,8-diol-9,10-epoxide (BPDE), which is highly reactive and binds covalently to DNA, leading to mutations and/or cancer (Figure 2) [19]. BP is sequentially metabolized to benzo[*a*]pyrene-7,8-oxide (BP-7,8-oxide), to benzo[*a*]pyrene-7,8-diol (BP-7,8-diol), and lastly to BPDE at the Bay-region, and each class of metabolic intermediate is genotoxic and carcinogenic [19, 22]. The Bay-region is defined as a concave exterior region on some PAHs that have three phenyl rings in a non-linear arrangement [22], and this theory has been widely accepted, particularly with four- or five-membered ring PAHs.

CYPs such as CYP1A1, CYP1A2, CYP1B1, and CYP3A4, in addition to epoxide hydrolase (EH), are involved in BP oxidative metabolism [29-37]. Unlike wild-type mice, transgenic mice deficient in *CYP1A1* showed enhanced toxicity after oral BP exposure, indicating the involvement of *CYP1A1* in the detoxification of BP [36, 37]. Robinson *et al.* [38] investigated the effects of oral administration of BP in several strains of mice, classified as "responsive" (those capable of producing increased levels of CYP-mediated enzymes as a consequence of PAH exposure) or "nonresponsive" (those not highly responsive to producing increased levels of CYP-mediated enzymes as a consequence of PAH exposure). They showed that decreased survival of "nonresponsive" mice was associated with their inadequate ability to increase CYP-mediated enzymes, and that death appeared to be due to bone marrow depression (aplastic anemia and pancytopenia). Prostaglandin synthase (also called cyclooxygenase, COX)-dependent cooxidation of BP-7,8-diol to BPDE is reported in the human lung and other mammalian cells [19, 37, 39, 40]. The stereochemistry of the metabolic transformation of BP to diols and diolepoxides is an important component of this mechanism of action [19, 22].

BPDE can form both stable and unstable (so-called "depurinating") adducts with DNA, primarily with guanines in many species and organs, mediated by electrophilic carbonium ions. (+)-*anti*-BPDE forms [+ta]-BP-N^2-deoxyguanosine, which is a major adduct formed by the *trans* addition of N^2-deoxyguanosine to (+)-*anti*-BPDE (abbreviated as BPDE-dG adduct) (Figure 2) [19, 41]. BP adduct at the N6 position of deoxyadenosine (N^6-BPDE-dA) is also detected [42, 43, 44].

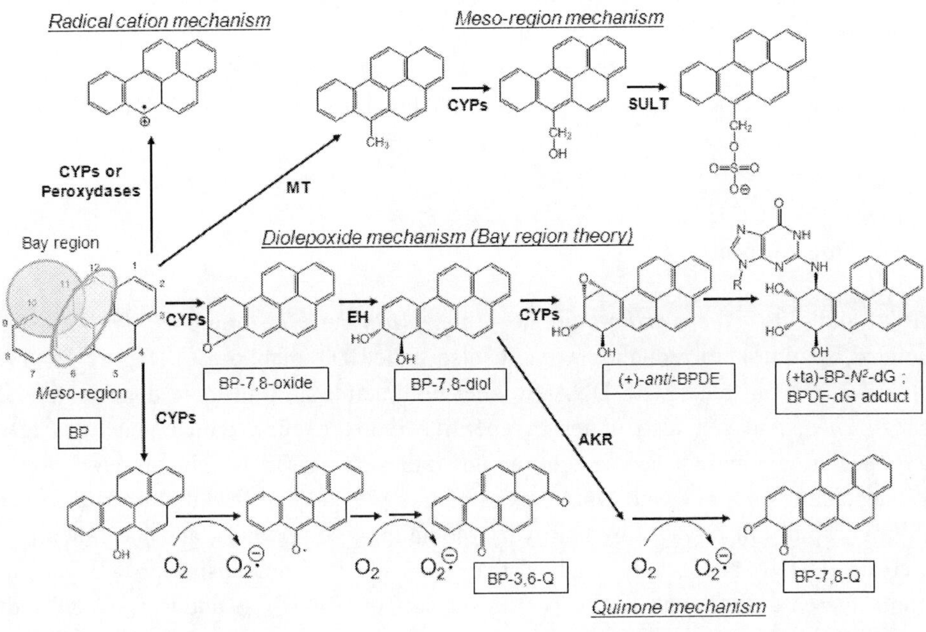

Figure 2. Pathway of metabolic activation from BP to various ultimate forms, which can react with DNA, leading to mutation/cancer. BP, benzo[*a*]pyrene; BP-7,8-oxide, benzo[*a*]pyrene-7,8-oxide; BP-7,8-diol, benzo[*a*]pyrene-7,8-diol; BPDE, benzo[*a*]pyrene-*trans*-7,8-diol-9,10-epoxide; BP-3,6-Q, benzo[*a*]pyrene-3,6-quinone; BP-7,8-Q, benzo[*a*]pyrene-7,8-quinone; EH, epoxide hydrolase; AKR, aldo-keto reductase; MT, methyltransferase; SULT, sulfotransferase.

BP orally administered to mice or rats does not seem to exhibit large differences in its abilities to form the major BPDE-dG adducts within a short period after exposure in various tissues, including forestomach, liver, spleen, and lung, in spite of the different animal species and strains [45-47]. The BPDE-dG adduct levels in different tissues from *Xpa*-deficient mice that lack the ability of DNA nucleotide excision repair following oral treatment for 7 weeks (13 mg/kg per treatment, 3 times per week) were in the range between 26 and 146 adducts/10^8 nucleotides in the 7 organs examined, with the smallest adduct level being in the colon, possibly due to a fast decrease of the DNA adduct *via* its high rate of cell division [48]. In *Apc*$^{Min/+}$ mice, which have a point mutation at the recessive *Apc* tumor suppressor gene [49, 50], following BP-treatment (a single *i.p.*, 100 mg/kg), the BPDE-dG adducts were found in adenomas ($56 \pm 15/10^8$ nucleotides) as well as in uninvolved small intestinal mucosa ($89 \pm 19/10^8$ nucleotides) [51], with both tissues having metabolic activation capacity [52]. The total DNA binding levels were also roughly equal (within several-fold) in some different organs, including the colon of BP-treated mice [53, 54]. The BPDE-dG adduct was reported to decay time-dependently at relatively slow rates with a half-life of 18 days in the lung and 9 days in the liver after oral treatment of A/HeJ mice with BP (6 mg/mouse) [45].

The DNA adducts may be converted into mutations through erroneous translesion sysnthesis by Y-class DNA polymerases (*e.g.,* DNA pol κ) during DNA replication [55]. The BPDE-dG adducts can induce base substitutions, frameshifts, insertions and deletions in *Escherichia coli,* and form mutations predominantly at the GC base pairs. GC → TA (57%), GC → AT (23%), and GC → CG (20%) mutations were each significant, and the nature of the mutation can be influenced by the sequence context of the adduct [41,56]. Tumors induced by BP had mutations in the oncogenes (*e.g.*, *ras* genes) or suppressor genes (*e.g.*, *p53* genes) [4]. Lung tumors were reported to have GC → TA and GC → AT mutations in the K_i-*ras* gene at codon 12 [57], and skin tumors were shown to have GC → TA at codon 13 and AT → TA mutations at codon 61 of the Ha-*ras* gene [58].

Quinone Mechanism

BP-7,8-diol to the *ortho*-quinone, benzo[*a*]pyrene-7,8-quinone (BP-7,8-Q), is metabolized by aldo-keto reductases (AKR, also called DT-diaphorase) (Figure 2) [19, 37, 59-61]. BP-7,8-Q can react with DNA to yield both stable and unstable depurinating DNA adducts *in vitro*, and can also undergo repetitive redox cycling which generates reactive oxygen species, including superoxide anion radicals ($\bullet O_2^-$), which produce 8-oxo-2'-deoxyguanosine (8-oxo-dG) and DNA strand-breaks [59, 60]. In the lung or livers of mice, BP-7,8-Q is rapidly inactivated by sulfhydryl-containing compounds, and inefficiently form stable covalent DNA adducts by Michael reaction with deoxyguanosine or deoxyadenosine [61], and thereby, the contribution of the resultant BP-DNA adducts to BP-induced tumorigenesis in mice is perhaps small. Comet test (also known as the Single Cell Gel Electrophoresis Assay) showed positive results in the liver and lung of Balb/c mice after a single dose of BP (125 mg/kg, *p.o.*) [62], indicating the production of single-stranded DNA or alkali-labile apurinic sites *in vivo*. In this study, 8-oxo-dG, which is an oxidative DNA damage, was also detected. Oxidative damage to DNA, produced by BP treatment, will be associated with this "quinone" mechanism.

BP is also known to be metabolized to 6-hydroxybenzo[*a*]pyrene (6-OH-BP) by CYPs, followed by spontaneous or enzymatical oxidation to benzo[*a*]pyrene-1,6-quinone, benzo[*a*]pyrene-3,6-quinone (BP-3,6-Q), and benzo[*a*]pyrene-6,12-quinone (Figure 2) [37]. These quinone derivatives are possible ultimate mutagens and/or carcinogens, and are detoxified by NAD(P)H quinone oxidoreductase (NQO1, formerly referred to as DT-diaphorase). The disruption of NQO1 in mice causes a decrease in BP-induced skin carcinogenicity [63].

Radical-Cation Mechanism

The radical-cation mechanism for BP has been studied in connection with mouse-skin tumorigenesis [19, 64]. One-electron oxidation of BP by CYPs or peroxidases such as prostaglandin H synthase creates a radical cation localized on carbon 6, as a consequence of the ionization potential and geometric configuration (Figure 2) [64]. In mouse skin, this radical cation gives rise to the formation of covalent adducts with guanine at the C8 carbon and N7 nitrogen, and adenine at the N7 nitrogen [20, 64]. These adducts are unstable and are considered to generate apurinic sites in mouse skin. Consequently, only low levels of apurinic sites were measured in the epidermis of mice treated with BP [65].

Meso-Region Mechanism

BP to 6-methylbenzo[*a*]pyrene, where the reactive *meso*- or L-region of BP is methylated, is transformed with *S*-adenosylmethionine as the carbon donor (Figure 2) [19, 66]. 6-Methylbenzo[*a*]pyrene is further metabolized by CYPs to 6-hydroxymethylbenzo[*a*]pyrene, and then conjugated to sulfate by sulfotranferase to 6-[(sulfooxy)methyl]benzo[*a*]pyrene. This reactive sulfate ester forms DNA adducts *in vivo* [67].

Aryl Hydrocarbon-Receptor Mechanism

The AhR regulates the transcription of a series of genes, including *Cyp1A1*, *Cyp1A2*, *Nqo1* (encoding NAD(P)H: quinine oxidoreductase-1), *Aldh3a1* (aldehyde dehydrogenase 3A1), *UGT1a6* (uridine 5'-diphosphate-glucuronosyl transferase 1A6), and *Gsta1* (glutathione *S*-transferase A1) [19, 37]. All these genes are activated by AhR-ligands *via* the AhR-mediated aromatic hydrocarbon response element, and BP can function as an AhR ligand. Knockout mice with AhR do not develop any tumors with BP, indicating that AhR is largely involved in BP carcinogenesis [68].

The AhR is also shown to play a role in the response to oxidative stress in cell-cycle regulation and apoptosis, which might modulate BP-carcinogenesis [69-71]. BP induces apoptosis by membrane remodeling following the reduced expression of 3-hydroxy-3-methylglutaryl-CoA reductase (HMG-CoA reductase), which is associated with AhR and BP-related H_2O_2 formation [69].

Epigenetic Mechanisms

Epigenetic alterations result in inheritable changes in gene expression and chromatin structure over many cell generations without changes in DNA sequence, leading to consequences equivalent to those induced by genetic alterations, and they include DNA methylation and histone (chromatin) modifications [72]. The CpG dinucleotide sequence, 60 to 90% of which contain 5-methylcytosine in human DNA, is of special significance in mammalian DNA, and abnormal hypo- or hyper-methylation is suggested to be associated with carcinogenesis in many different kinds of organs/tissues [72, 78, 79]. Methylated CpG sites not only greatly increase BP-DNA adduct formation at CpG sites, but also affect BP-DNA adduct formation at surrounding sequences, at least in *in vitro* studies using oligonucleotides [44,74,75]. The *p53*-specific disruption of the cell cycle and the disruption of DNA methylation at CpG sites were reported in breast cancer cells following BP treatment [73].

The H2A histone variant H2AX is an important factor involved in the process of safeguarding genomic stability, and H2AX undergoes a specific phosphorylation at serine 139 at the C-terminal, denoted as γH2X for efficient DNA repair upon formation of DNA double-stranded breaks [76-78]. H2AX phosphorylation is induced in human lung epithelial carcinoma cells following BPDE-treatment, suggesting that resultant chromatin remodeling may also play a critical role in setting the overall genotoxicity and carcinogenicity of BP [80].

BP can induce changes in the expression of many nuclear proteins, several of which are involved in the alternative splicing of mRNA, DNA replication, recombination and repair in human HeLa cells [81]. Many of the reactive intermediates of BP (oxides, diolepoxides, quinones, and radical cations), and repetitive redox cycling-generated reactive oxygen species (Figure 2), can disrupt the balance of cellular oxidants and anti-oxidants, resulting in alteration of cell signaling.

Immunosuppression Mechanism

BP and some BP metabolites have the ability to suppress the immune system. Transgenic mice deficient in *CYP1A1* or *CYP1B1* showed toxicity in the spleen or bone marrow, resulting in immunosuppression after BP exposure [19, 36, 37]. Furthermore, BP suppresses immunity by *p53*-dependent pathways by modulating signaling pathways in lymphocytes *via* non-genotoxic mechanisms, and by oxidative stress [19, 82]. Metabolites mediated by CYP1A1 suppress the growth of human B cells, leading to immunosuppression [83].

Current Evaluation for the Carcinogenic Risk of BP by Food-Intake

The margin-of-exposure (MOE) approach has recently been recommended by both the FAO/WHO joint Expert Committee on Food Additives (JECFA) and the European Food safety Authority for the assessment of substances in food that are both genotoxic and carcinogenic [84,85]. The approach involves comparison of the magnitude of a selected (lower) point on the dose-response curve (*e.g.,* BMDL10 which is the estimate of the lower 95% confidence interval of the Bench Mark Dose that produces a 10% increase in the incidence of a particular tumor over background), termed the point of departure, with the human exposure estimate(s) to the carcinogen. JECFA has concluded that an MOE of 10,000 and above would be of little concern from a public health point of view, based on a BMDL10 drawn from an animal study and the overall uncertainty factors [84, 85].

BP was applied to the MOE approach [26]. The BMDL10 of BP was estimated as 0.122 mg/kg/day, based on the number of total tumor-bearing mice following oral exposure (in the diet) to coal tar mixture that contained BP for 2 years. An average dietary exposure estimate was 8 ng/kg/day. Thus, the calculated MOE was 15,000, estimating that BP would be of little concern.

STUDIES ON THE GENOTOXICITY AND CARCINOGENICITY OF BP

Like the other PAHs, BP has not received so much attention as a cause of environmental chemical-induced human CRCs, because all the studies using normal rodents failed to show clear evidence of the carcinogenic potential of BP [1, 19, 23, 25, 26]. Moreover, the number of animal experiments of BP for carcinogenicity induced by oral exposure, is very limited. Therefore, we have comprehensively studied the genotoxicity and carcinogenicity of BP, particularly focusing on the murine colon. Our studies presented in this chapter are largely divided into: (1) studies on genotoxicity and carcinogenicity in mice, and (2) study using *Salmonella* mutagenicity test (Ames test).

A series of studies using mice is outlined in Figure 3; *in vivo* mutagenicity including mutation spectrum, *in vivo* clastogenicity, carcinogenicity, cell proliferation rate, hematology and clinical chemistry, were examined following oral administration of BP at 75 or 125 mg/kg/day for 5 consecutive days [86-89]. Muta^{TM}Mouse (strain 40.6) or mouse with the same genetic background strain, CD2F$_1$ (BALB/c x DBA2) mouse, was used throughout a series of our studies, except for ICR mouse to examine the cell proliferation rate. MutaMouse was originally developed for use in *in vivo* mutation assay, and this transgenic mouse carries approximately 80 copies of *E. coli lacZ* genes with 3.1 k base pairs in every cell, encoding *E. coli* β-galactosidase, as a marker gene for the detection of mutations [90-95].

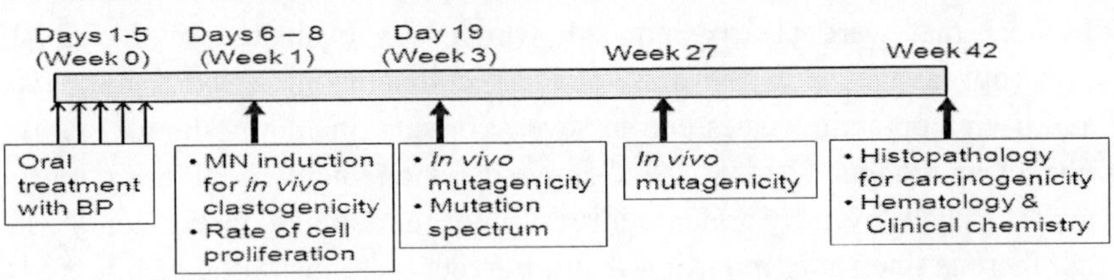

Figure 3. Outline of a series of our studies to comprehensively examine the genotoxicity and carcinogenicity of BP in male Muta^{TM}Mouse (CD2F$_1$ mice) after oral treatment with BP at 75 and/or 125 mg/kg/day or vehicle (corn oil) for 5 consecutive days (from Days 1 to 5). Genotoxicity in MutaMouse was evaluated by micronucleus induction (micronucleus test) at Days 6 and 7, when the rate of induction of micronucleated reticulocytes in peripheral blood was determined, and mutation induction in *in vivo* mutagenicity tests at Weeks 3 and 27, when the rate of induction of mutated *E. coli lacZ* transgene was determined. Mutation spectrum of mutated *lacZ* transgene in MutaMouse was determined using the same organs used to determine the rate of mutation induction at Week 3. MutaMouse was autopsied for histopathology to examine the carcinogenicity of BP, and for hematology and clinical chemistry. Rate of cell proliferation by the BrdU method was also examined in male ICR mice at Days 6 and 8 after oral treatment with BP at 125 mg/kg/day or vehicle (corn oil) for 5 consecutive days.

Carcinogenicity of BP Induced by Oral Administration

(1) Carcinogenicity of BP in MutaMouse

The carcinogenic potency of BP in MutaMouse was examined 41 weeks (at Week 42) after administration (Figure 3) [86, 87]. Two of 12 mice exposed to 125 mg/kg/day BP were euthanized *in extremis* at Weeks 27 and 40 due to malignant lymphoma. Moreover, one of 10 mice was found dead at Week 41, and the remaining mice showed poor clinical conditions at Week 42.

Therefore, autopsy was carried out for the analysis of histopathology, hematology and clinical chemistry at that week.

Hiatopathological analysis showed squamous cell carcinoma, papilloma, and hyperplasia in the forestomach, malignant lymphoma in the lymphatic organs including spleen, and bronchiolar-alveolar hyperplasia in the lung (Table 1). Thus, the target organs of BP-induced carcinogenesis in MutaMouse by oral treatment are forestomach, lymphatic organs including spleen (lymphoma) and lung, whereas colon is not a target organ. MutaMouse necropsied at Week 42 were in poor clinical conditions; however, these mice showed only slight anemia (Table 2). Two mice with malignant lymphoma, each necropsied at Weeks 27 and 40, showed increased count of white blood cells and severe anemia (Table 2) [87]. Only a slight change in some parameters of clinical chemistry was observed in mice necropsied at Week 42 (Table 3) [87].

(2) Carcinogenicity of BP in Normal Animals

IARC monographs [19] documented that BP produces tumors in many species (mouse, rat, hamster, guinea-pig, rabbit, duck, newt, and monkey) following exposure by many different routes (oral, dermal, inhalation, intratracheal, intrabronchial, subcutaneous, intraperitoneal, and intravenous). BP had both local and systemic carcinogenic effects [19]. Table 4 summarizes the carcinogenicity of BP in mice, rats and hamsters, as target organs for malignant tumors that were clearly induced, which was evaluated by IARC [19]. Oral administration (by gavage or in the diet) of BP resulted in increased tumor responses in lymphoid and haematopoietic tissues and in several organs, including the lung, forestomach, liver, oesophagus and tongue [19, 25, 26]. Colonic carcinogenicity of BP has not been clearly shown in normal rodents even by oral administration or intracolonic instillation [1, 19, 23, 25, 26], although there is one study reporting that intrarectal administration of 0.8 mg BP/kg for 30 weeks produced adenocarcinomas in the colon of Syrian Hamsters [96].

(3) Carcinogenicity of BP in Transgenic Mice

BP given orally to transgenic mice that lack the ability of DNA nucleotide excision repair (*XPA*) or weaken the function to suppress carcinogenesis (*p53*), induced tumors much earlier and at higher incidences than similarly treated wild-type mice. Namely, lymphoma was induced in $XPA^{-/-}$ mice, and primarily lymphomas and forestomach tumors were induced in $XPA^{-/-}/p53^{+/-}$ double-transgenic mice [97, 98]. Moreover, XPA knock-out mice are shown to develop lymphomas after oral treatment [99], or pulmonary tumors after intratracheal instillation [100] with BP. Transgenic mice which lung-specifically express the dominant-negative mutant *p53* had their lung cancer increase from 39% (wild-type mice) to 73% by 6 months after beginning intratracheal instillation of BP [101].

Table 1. Incidence of tumors and hyperplasias in MutaMouse 26 (at Week 27) and 41 weeks (at Week 42) after oral administration of BP for 5 consecutive days

Group	BP (mg/kg/day)	No. of effective mice		Forestomach		Spleen	Lung
			Squamous cell carcinoma	Squamous cell papilloma	Squamous cell hyperplasia	Malignant lymphoma	Bronchiolar-alveolar hyperplasia
Week 27	0	4	0 (0)[a]	0 (0)	0 (0)	0 (0)	0 (0)
	125	4	0 (0)	1 (25)	2 (50)	0 (0)	0 (0)
Week 42	0	8	0 (0)	0 (0)	0 (0)	0 (0)	0 (0)
	75	11	2 (18)	4* (36)	10* (91)	0 (0)	2 (18)
	125	11	2 (18)	5* (45)	10* (91)	2[b] (18)	1 (9)

[a]: Numbers in parentheses indicate percentage relative to vehicle control in each group.
[b]: Two mice were euthanized in extremis at Weeks 27 and 40.
*: $P < 0.05$ in a χ^2 test. Data are taken from our study [86].

Table 2. Hematological parameters[a] for MutaMouse 41 weeks (at Week 42) after oral treatment with BP at 125 mg/kg/day for 5 consecutive days

Dose (mg/kg/day)	No. of effective mice	Red blood cells (x 10^6/μL)			Hemoglobin (g/dL)			Hematocrit (%)			White blood cells (x 10^3/μL)			Platelet (x 10^3/μL)		
0	8	10.00	±	0.65	14.6	±	0.7	46.6	±	2.1	2.47	±	1.00	1158	±	82
75	11	8.50	±	1.90	12.5	±	2.7	40.1*	±	7.5	2.07	±	1.30	1293	±	226
125	9	8.26*	±	2.00	12.2*	±	2.9	40.2	±	7.9	2.54	±	1.82	1199	±	287
125	2[#]	4.17	±	2.28	5.85	±	3.0	19.2*	±	3.6	47.98	±	55.16	733	±	672

#: mice with malignant lymphoma, euthanized *in extremis* at Weeks 27 and 40, among 11 effective mice of the 125 mg/kg/day group.
[a]: mean ± standard deviations, *; $p < 0.05$ in Student's *t* test. Data are taken from our study [87].

Table 3. Clinical chemistry parameters for MutaMouse 41 weeks (at Week 42) after oral treatment with BP at 125 mg/kg/day for 5 consecutive days

		Dose (mg/kg/day)								
		0			75			125		
Parameter		(n = 8)			(n = 11)			(n = 11)[#]		
Total protein (TP)	(g/dL)	4.8	±	0.2	4.5	±	0.3	4.4	±	0.4*
Albumin (ALB)	(g/dL)	2.8	±	0.1	2.6	±	0.1	2.7	±	0.2
Total cholesterol (T-Chol)	(mg/dL)	69	±	8	68	±	9	74	±	26
Phospholipids (PL)	(mg/dL)	151	±	19	146	±	21	146	±	32
Triglycerides (TG)	(mg/dL)	53	±	21	75	±	32	102	±	107
Glucose (Glu)	(mg/dL)	198	±	22	192	±	13	217	±	36
Urea nitrogen (UN)	(mg/dL)	25.2	±	3.1	27.3	±	5.1	33.9	±	6.0**
Creatinine (Creat)	(mg/dL)	0.08	±	0.01	0.08	±	0.02	0.12	±	0.01**
Alkaline phosphate (AP)	(IU/L)	116	±	17	89	±	19*	101	±	36
Alanine aminotransferase (ALT)	(IU/L)	19	±	4	15	±	2*	20	±	15
Aspartase aminotransferase (AST)	(IU/L)	37	±	11	30	±	6	53	±	70[##]
Inorganic phosphate (IP)	(mg/dL)	7.3	±	1.9	6.1	±	1.2	6.9	±	2.1
Calcium (Ca)	(mg/dL)	8.2	±	0.1	7.9	±	0.4	8.6	±	1.3
Sodium ion (Na)	(mEq/L)	151	±	2	152	±	1	150	±	2
Potassium ion (K)	(mEq/L)	4.1	±	0.4	4.4	±	0.4	5.3	±	0.7**
Chloride ion (Cl)	(mEq/L)	115	±	3	118	±	1	114	±	2

Numbers in parentheses indicate those of effective mice.

[#]: including 2 mice with malignant lymphoma, euthanized *in extremis* at Weeks 27 and 40.

[##]: Five-fold increases over the mean value of the vehicle group in two mice with malignant lymphoma might be due to their metastasis to other organs including the liver.

Mean ± standard deviations, *: $p < 0.05$, **: < 0.01 in Student's *t* test.

Data are taken from our study [87].

Table 4. Malignant tumors clearly induced in experimental animals by BP, evaluated by IARC [19]

	Lung	Trachea	Larynx	Forestomach	Liver	Lymphoid tissue (lymphoma)	Sarcoma (Injection site)	Skin	Mammary gland
Mouse	x			x	x	x	x	x	
Rat	x						x		x
Hamster	x	x	x	x			x		

In E mu-pim-1 transgenic mice that are predisposed to develop lymphomas, lymphomas were increased following oral treatment with BP 3 times a week for 13 weeks up to a 90% incidence [102]. Skin tumorigenicity was shown in transgenic TG-AC mice carrying the oncogenic v-Ha-*ras* 25 weeks after treatment with 40 μg of BP twice a week at a 100% tumor response [103]. Recently, the carcinogenic potency of BP was examined in *Apc*^{Min} mice, which have a point mutation at the recessive *Apc* tumor suppressor gene [49, 50]. Most of the polyps observed in the colon of BP-treated *Apc*^{Min} mice (50 and 100 μg/kg BP daily for 60 days orally) were adenomatous polyps with low- to high-grade dysplasia [104]. The transgenic *Apc*$^{Min+}$ and CYP1b1 deficient mice (*Apc*$^{Min/+}$, CYP1b1$^{-/-}$) administered with BP (0.1 mg/kg on days 28 and 35 intraperitoneally) presented adenomatous polyps, and the mice that lacked CYP1b1 presented few tumors compared to the ones that had CYP1b1 [105].

In Vivo Genotoxicity of BP Induced by Oral Administration

(1) Micronucleated Reticulocytes Induced by BP in MutaMouse

Micronucleus tests of BP were conducted 24 hours (at Day 6) and/or 48 hours (at Day 7) after oral treatment of MutaMouse with BP at 125 mg/kg/day for 5 consecutive days to examine the clastogenic potency (Figure 3). In this test, the frequency of micronucleated reticulocytes (MNRETs) per 2000 reticulocytes (RETs) was determined from peripheral blood. The frequencies of MNRETs following BP treatment were increased by 4 to 5 folds as compared to vehicle (corn oil) treatment (Table 5, our unpublished data). This result indicates that BP induces chromosome damage in mouse bone marrow cells after oral treatment, consistent with a previous report [106].

Table 5. MNRETs in peripheral blood of MutaMouse[a] following oral treatment with BP at 125 mg/kg/day for 5 consective days

Dose (mg/kg/days)	Sampling time (hours)[b]	No. of MNRETs/2000 RETs (%)		
0	48 (Day 7)	0.25	±	0.06
75	48 (Day 7)	1.05**	±	0.45
125	24 (Day 6)	1.05**	±	0.20
125	48 (Day 7)	1.20**	±	0.26

[a]: Four mice were examined for micronucleus induction.
[b]: hours after the last dosing.
RETs; reticulocytes, MNRETs; micronucleated reticulocytes.
**: $p < 0.01$ in Student's *t* test.

(2) Mutation Frequency Induced by BP in MutaMouse at Week 3

BP was examined for *in vivo* mutagenicity in MutaMouse (Figure 3) [86]. Two weeks (at week 3) after administration of BP, a marked increase in the frequencies of *lacZ* mutations was found in multiple organs including the 3 target organs for BP-carcinogenesis (Figure 4).

However, the magnitude of *lacZ* mutation frequencies induced by BP in different organs did not necessarily associate with the target organs for carcinogenicity. More interestingly, the colon, which was a non target organ for BP-carcinogenesis, was the organ that showed the highest increase in mutation frequency (a 37-fold increase over the spontaneous frequency) of the 11 organs examined.

Kosinska *et al.* [107] reported that mutations were induced in the mammary and oral tissues, in addition to the colon of MutaMouse orally exposed to BP. Following oral BP treatment, spleen, liver, and lung were also shown to be mutated in both mice harboring *lacZ* reporter genes, which were proficient ($XPA^{+/+}$/*lacZ*) or deficient ($XPA^{-/-}$/*lacZ*) in nucleotide excision repair genes [108].

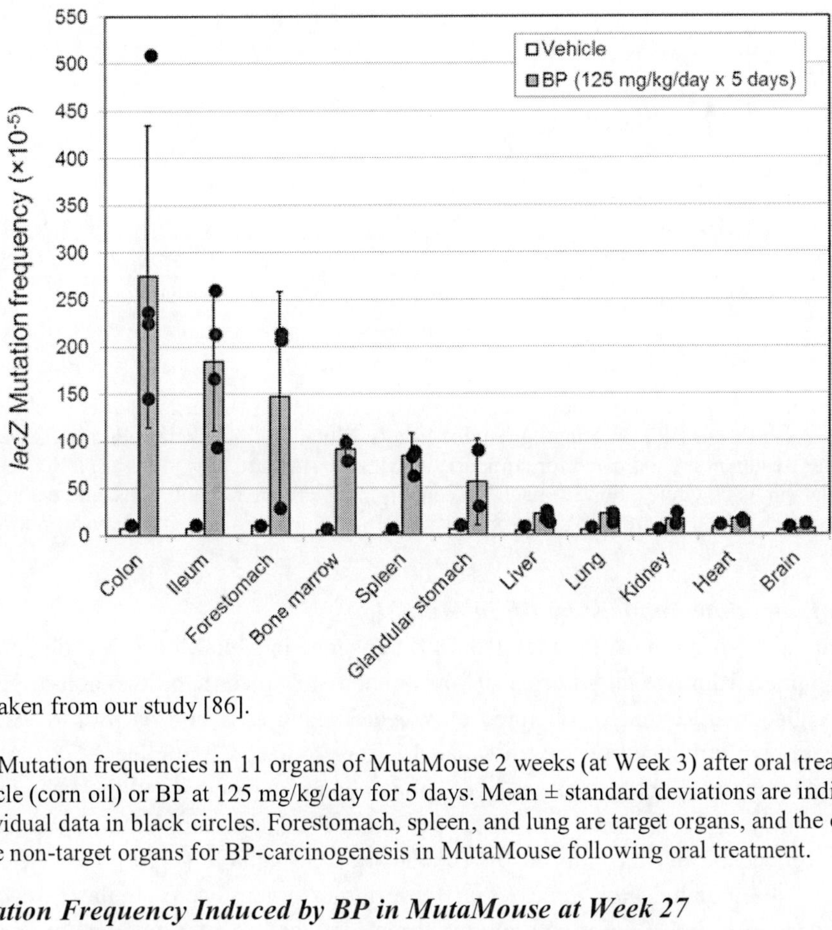

Data are taken from our study [86].

Figure 4. Mutation frequencies in 11 organs of MutaMouse 2 weeks (at Week 3) after oral treatment with vehicle (corn oil) or BP at 125 mg/kg/day for 5 days. Mean ± standard deviations are indicated with individual data in black circles. Forestomach, spleen, and lung are target organs, and the other organs are non-target organs for BP-carcinogenesis in MutaMouse following oral treatment.

(3) Mutation Frequency Induced by BP in MutaMouse at Week 27

The frequency of mutations in MutaMouse was also determined in 3 target organs (forestomach, spleen, and lung) and 3 non-target organs (colon, glandular stomach, and liver) 26 weeks (at Week 27) after administration (Figure 5) [88]. Mutation frequencies induced in these organs, including colon at Week 27, were equal or slightly higher than those at Week 3. This finding indicates that cells carrying the *lacZ* transgene mutated by BP remain in each organ for a long period of time, at least 26 weeks after treatment. If DNA lesions had been produced in differentiated colonic epithelial cells following BP exposure, mutations would not have been induced. This is because these cells have no ability to divide (cell division is necessary for a mutation to fix), and they fall out of the colonic mucosa through apoptosis in several days. Therefore, it is considered that mutations probably occurred in either the colonic stem cells or progenitor cells, which rapidly divide [4, 109-111]. In any case, the colon is highly mutated in mice following oral exposure of BP.

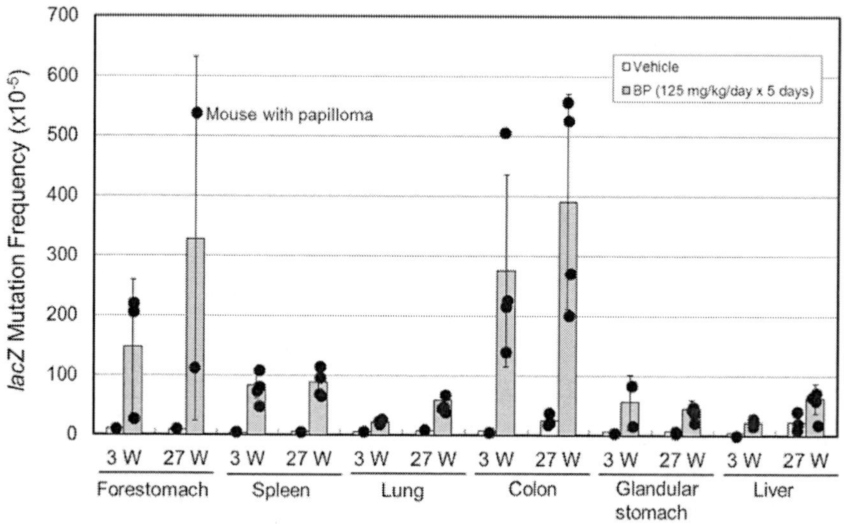

Data are taken from our study [88].

Figure 5. Mutation frequencies in 6 organs of MutaMouse 26 weeks (at Week 27) after oral treatment with vehicle (corn oil) or BP at 125 mg/kg/day for 5 days. Mutation frequencies observed 2 weeks (at Week 3) after treatment are indicated for comparison. Mean ± standard deviations are indicated with individual data in black circles. Forestomach, spleen, and lung are the target organs, and colon, glandular stomach, and liver are non-target organs for BP-carcinogenesis in MutaMouse following oral treatment.

(4) Type of mutations Induced by BP in MutaMouse

The mutation spectrum of the mutated *lacZ* transgene in MutaMouse was determined for samples obtained from two target organs (forestomach and spleen) and two non-target organs (colon and glandular stomach), obtained at Week 3 (Figures 3 and 4) [89]. These organs selected were highly mutated by BP; they had 15-(forestomach), 25-(spleen), 37-(colon), and 9.2-(glandular stomach) fold increases over spontaneous frequency with the vehicle (Figure 4). Hence, analysis of the mutation spectrum for BP-treated samples was considered to practically reflect the events induced by BP, but not the vehicle (Figure 6). The mutation spectrums of the mutated *lacZ* genes from these target organs were similar to each other. GC→TA base pair substitution mutations (transversions) were predominant, presenting approximately half of all the mutations. In contrast, the mutation spectrum of the mutated *lacZ* genes from these non-target organs were not similar to each other, and, more importantly, were different from those obtained from these target organs (colon; a decrease in GC→TA and increases in GC→CG and GC→AT, and glandular stomach; a decrease in deletion and an increase in GC→AT, as compared to these target organs). These findings suggest that the mutation spectrum may be important for the induction or development of BP-carcinogenesis, because it may affect activation or deactivation of the gene expression of the genes involved, such as oncogenes or tumor suppressor genes in each organ.

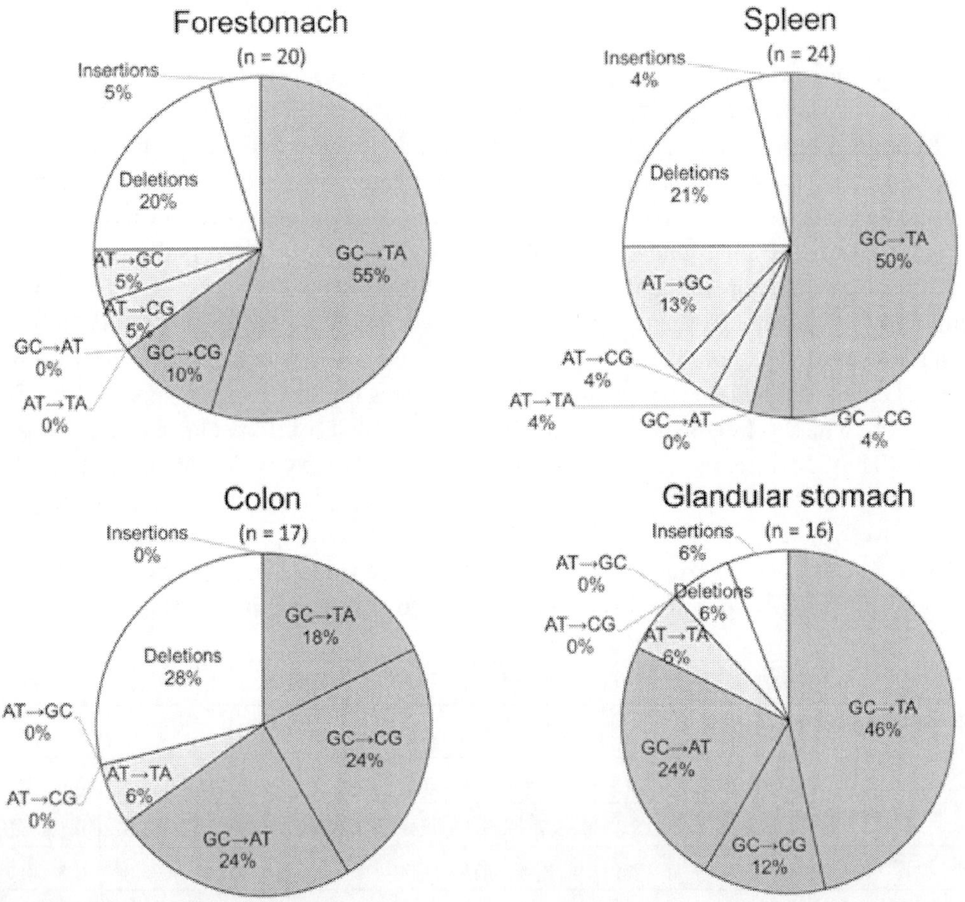

Data are taken from tour study [89].

Figure 6. Mutation spectrum of mutated *lacZ tr*ansgene from four organs of MutaMouse at Week 3. Forestomach and spleen are target organs, and colon and glandular stomach are not for BP-carcinogenesis.

Predominant mutations produced by BP in mice were base pair substitutions (72%-88% of all types of mutations, depending on the tissue examined). This finding was consistent with other studies reporting the mutation spectrum in the liver of MutaMouse orally exposed to BP [112], and in the lung of *gpt* delta transgenic mice (mice capable of *gpt* mutation) intratracheally instilled with BP [113]. These findings were both roughly similar to the mutation spectrum found in the glandular stomach in this study [89].

Molecular epidemiological studies suggest that BP, which is generated from smoking and incomplete combustion of coal, is attributable to lung cancers. Robles *et al.* [114] reported that GC→TA transversions have been found in 15% of a group of more than 15,000 mutant *p53* alleles associated with a variety of human tumors. However, 21% of mutant *p53* alleles carried this transversion in lung carcinomas from nonsmokers, 33% carried it in cigarette smokers, and 75% carried it in nonsmokers who had a history of repeated exposure to smoky coal emissions.

Cell Proliferation in Mice after Oral Administration of BP

Cell proliferation is known to play a critical role in carcinogenesis [4, 115-117]. Therefore, we determined the rate of cell proliferation in several organs 24 (Day 6) or 72 (Day 8) hours after oral treatment of ICR mice with a vehicle or BP at 125 mg/kg/day for 5 days (Figure 3). The organs examined included the 3 target organs (forestomach, spleen, and lung) and 2 non-target organs (colon and liver). Bromodeoxyuridine (BrdU) was used to label the colonic epithelial cells undergoing DNA replication [4]. Of the organs examined, the target organ of forestomach only enhanced cell proliferation, whereas the rates of cell proliferation of the other organs were not enhanced following BP exposure (Table 6, our unpublished data). The colon's rate of cell proliferation was not enhanced by BP treatment. However, colon (epithelial cells) intrinsically proliferates rapidly; its spontaneous rate of cell proliferation without treatment is relatively equal to that of forestomach following BP treatment. Thus, such a high rate of cell proliferation may carry with it the potential risk of colonic cancer.

Table 6. Cell proliferation in 5 organs of ICR mice treated orally with BP at 125 mg/kg/day for 5 consecutive days (by the BrdU method)

Dose	Sampling time	Forestomach	Spleen	Lung	Colon	Liver
(mg/kg/day)	(hours)[a]			(%)		
0	24 (Day 6)	6.9 ± 3.6	0.8 ± 0.4	0.0 ± 0.0	16.4 ± 6.3	0.6 ± 0.5
125	24 (Day 6)	25.2** ± 5.7	1.1 ± 0.4	0.0 ± 0.0	17.7 ± 1.3	0.1 ± 0.0
125	72 (Day 8)	12.4* ± 4.8	1.5 ± 0.7	0.0 ± 0.0	17.7 ± 2.8	0.5 ± 0.7

[a]: hours after the last dosing.
*: $p < 0.05$, **: $p < 0.01$ in Student's t test.

Induction of Colonic Cancer by BP in a Dextran Sulfate Sodium (DSS)-Induced Colon Colitis-Associated Model

(1) DSS-Induced Colon Colitis Model

These days, colitis has become widely accepted as closely associated with colonic cancer in humans and animals [2, 4, 118-121]. Recent studies show that 1-4 cyclic administration of DSS, a non-genotoxic sulfated polysaccharide, in drinking water (each cycle; 3-7 days of DSS followed by untreated water for recovery) at a concentration ranging from 1-5% to mice, results in the establishment of chronic colitis and the development of colonic tumors with pathological features that resemble those of human colitis-associated colon cancer [2, 119, 120, 122-124]. However, this model provides a low incidence of colonic tumors and requires a long time for tumor formation. For example, Swiss Webster mice after 4 cycles of treatment, with each cycle consisting of 1 week of DSS followed by 2 weeks of untreated water, developed adenocarcinomas in 15-20% of mice, but required a minimum of 12 weeks to do so [122].

(2) DSS-Induced Colon Cancer Models in Combination with Colon Carcinogens

To enhance tumor incidence and multiplicity, the DSS regimen has been added with a colon carcinogen in normal mice or *Apc*$^{Min/+}$ mice to develop accelerated mouse colon cancer models [2, 120, 125-132].

DSS models modified by combining them with a colon carcinogen consists of 2 steps: 1) treatment of a methylating agent such as azoxymethane (AOM) and 1,2-dimethylhydrazine, or food-derived HCAs as an initiator for colon carcinogenesis, and 2) subsequent DSS treatment to induce colitis as a potent promoter, probably through direct epithelial toxicity and acute/chronic colonic inflammation [2, 118, 120, 121].

(3) DSS-Induced Colon Cancer Models in Combination with BP

We recently applied BP to the DSS-induced colon colitis model (Figure 7) [133]. Male $CD2F_1$ mice with the same genetic background as MutaMouse were orally administered BP at 125 mg/kg/day for 5 days (at Week 0), and starting 10 days after the last treatment, given 4% DSS in drinking water for 1 week (at Week 2).

At Week 4, multiple colonic adenomas were found in 100% of mice, and colonic adenocarcinomas were observed in 5/8 mice at a multiplicity of 1.5 ± 1.8 (per mouse) (Table 7). Mice administered BP and subsequent DSS for 1 week twice (intermittently, at Weeks 2 and 5 separated by a 2-week DSS treatment free

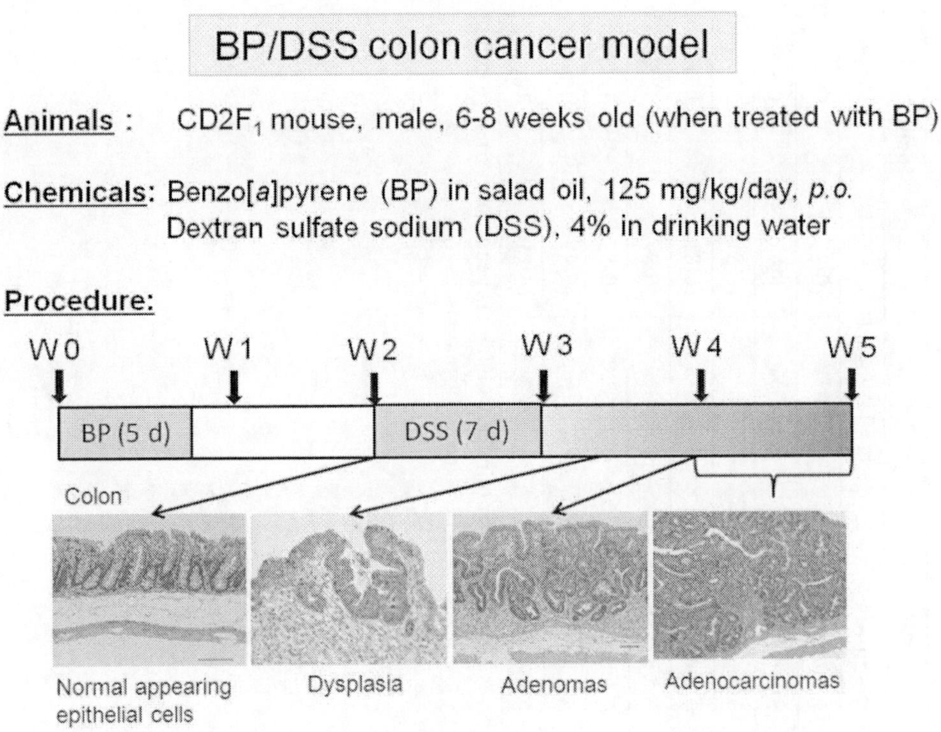

Figure 7. Outline of the BP/DSS colon cancer model that we have recently developed.

Table 7. Incidence and multiplicity of colonic neoplasia found in CD2F$_1$ mice

		Incidence*				Multiplicity**	
Group	**Diagnosis**	**Week 4**		**Week 11**		**Week 4**	**Week 11**
Vehicle	Adenoma	0%	(0/8)	0%	(0/8)	0	0
	Adenocarcinoma	0%	(0/8)	0%	(0/8)	0	0
BP	Adenoma	0%	(0/8)	0%	(0/8)	0	0
	Adenocarcinoma	0%	(0/8)	0%	(0/8)	0	0
DSS	Adenoma	0%	(0/8)	12.5%	(1/8)	0	0.1 ± 0.4
	Adenocarcinoma	0%	(0/8)	12.5%	(1/8)	0	0.3 ± 0.7
BP/DSS	Adenoma	100%	(8/8)	100%	(6/6)	5.9 ± 2.5	2.8 ± 1.0
	Adenocarcinoma	62.5%	(5/8)	100%	(6/6)	1.5 ± 1.8	23.3 ± 7.0

*: figures in parentheses indicate number of mice with tumors per number of mice examined.

**: number of tumors per mouse, mean ± standard deviations.

Data are taken from our study [133].

interval period) showed colonic adenocarcinomas in 100% of mice and at a high multiplicity of 23.3 ± 7.0 at Week 11 (Table 7). At this time point, multiple large masses could be macroscopically detected (Figure 8). Histopathological findings observed in this novel model showed similarities to those reported in other DSS-induced colitis-associated colon rodent models in combination with and without a colon carcinogen; tubular adenomas and well- or moderately-differentiated adenocarcinomas [2, 119, 120, 122-132]. In contrast, adenoma and adenocarcinomas were observed only in 1/8 mice treated with DSS alone at Week 11, but no proliferative lesions were observed in mice exposed to BP alone or vehicle (salad oil). These results suggest that BP is a possible colon carcinogen in humans with colitis, and that inflammation induced by DSS in addition to mutagenesis by BP at least, in part, contribute to colon carcinogenesis.

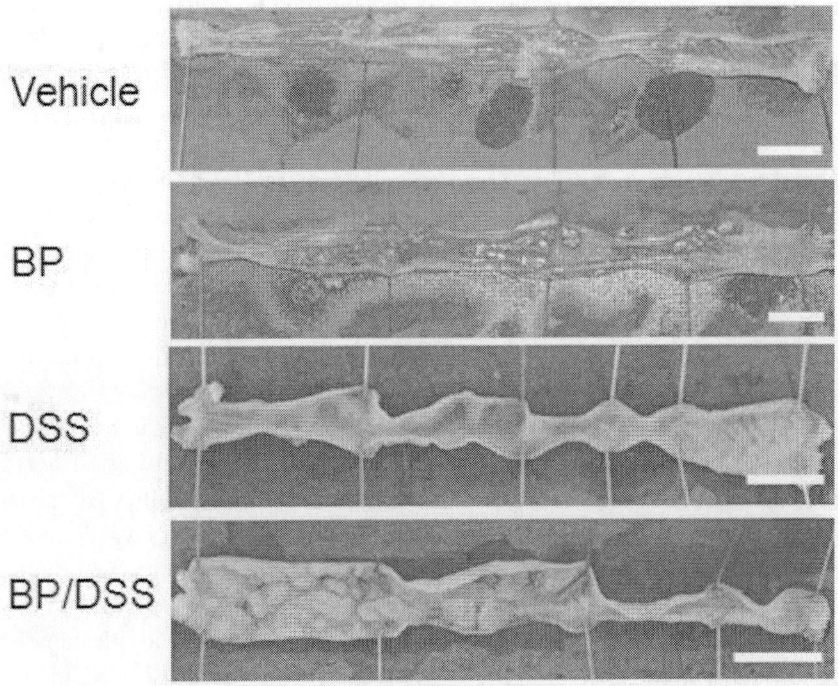

Data are taken from our study [133].

Figure 8. Macroscopic view of large bowels obtained at Week 11 from mice exposed to either vehicle, BP, DSS, or BP and DSS in combination. See experimental procedure for the BP/DSS model, shown in Figure 7. Bar: 1 cm.

Colonic tumors were observed at as early as Week 4 in this study, implying that they were produced within 2 weeks after the initial treatment with DSS for 1 week following exposure to BP (Figure 7, Table 7). Therefore, the rapid induction of adenocarcinomas might be interpreted not by a multi-hit model (sequential genetic alterations), which has been proposed in many studies on CRC in human and animal models [2, 4, 49], but rather by a single-hit model, *i.e.,* a consequence of direct formation (without sequential genetic alterations) from BP-mutated colonic epithelial stem cells or progenitor cells into adenocarcinoma cells, through transiently-appearing dysplastic foci and subsequent adenomas after exposure to DSS. Our current speculations on the mechanisms for the tumorous transformations include (1) changes in crypt cell metabolism, closely associated with inflammation, (2) alterations of signal transduction, (3) disruption of microenvironment surrounding colonic crypt, (4) increased cryptal cell proliferation, accompanied with regenerative mucosa, and (5) epigenetic alterations such as abnormal DNA methylation and histone modifications [2, 4, 120,134-140].

Mutagenicity of BP in the Ames Test Using Human S9

Manifold studies have demonstrated that BP is genotoxic in different kinds of *in vitro* genotoxicity testing systems and different kinds of bacterial or mammalian cells in the presence of different kinds of S9 fractions (supernatant fraction obtained after centrifugation of tissue homogenate at 9,000 *g*) that are prepared from rodent livers [4, 19, 22]. The *Salmonella* mutagenicity test (Ames test) has been the most world-wide *in vitro* genotoxicity test [141, 142]. Many chemicals remain inactive until they are enzymatically transformed to electrophilic species capable of covalently binding to DNA, leading to mutation and/or cancer [29-37]. Therefore, S9 fractions that are prepared from rat liver are often added to the treatment mixture, fortified by an NADPH-generating system in genotoxicity tests including the Ames test [141, 142]. S9 fraction is usually prepared from rats pretreated with either phenobarbital/5,6-benzoflavone (rat S9 (PB/BF)) or Aroclor 1254 to enhance drug metabolizing enzyme activity for the detection of chemical mutagenicity [141, 142]. In contrast, human S9 fractions have usually not been used in genototoxicity studies [24, 143-149].

For a more comprehensive assessment of the mutagenic risk of BP in humans, including assessment of inter-individual variation, we examined the mutagenicity of BP in the Ames test using 18 human liver S9 fractions from individual donors [149]. BP produced an extraordinarily wide distribution of mutagenic response (Figure 9). Two donors (lots) of human S9 fractions, HLS-014 and HLS-059, showed potent mutagenicity, while the other 16 lots of human S9 fractions had weak or marginal mutagenicity. HLS-014 had one of the highest P450 enzyme activities, particularly CYP3A4 and CYP2D6 activities among the S9 fractions used, and a high level of these enzyme activities was considered to be enzyme induction by intake of anti-asthma agents for 10 years or genetic polymorphisms in drug-metabolizing enzymes. The mutagenicity of BP in the presence of HLS-014 was equal to that observed in the presence of rat S9 (PB/BF), and its mutagenicity in the presence of pooled human liver S9 fraction (lot. HLSP-E5) was 0.05-fold lower than that observed in the presence of rat S9 (PB/BF). Base-pair substitution mutations induced by BP, which can be detected in the *Salmonella* TA100 test strain [141, 142], are considered to be attributable to BPDE-dG adducts after metabolic activation of BP [22, 41]. Large inter-individual variation in the mutagenicity induced by BP may be consistent with the report showing a 30-fold variation in the BPDE-dG level, in spite of the small variation in the total metabolism of BP [150].

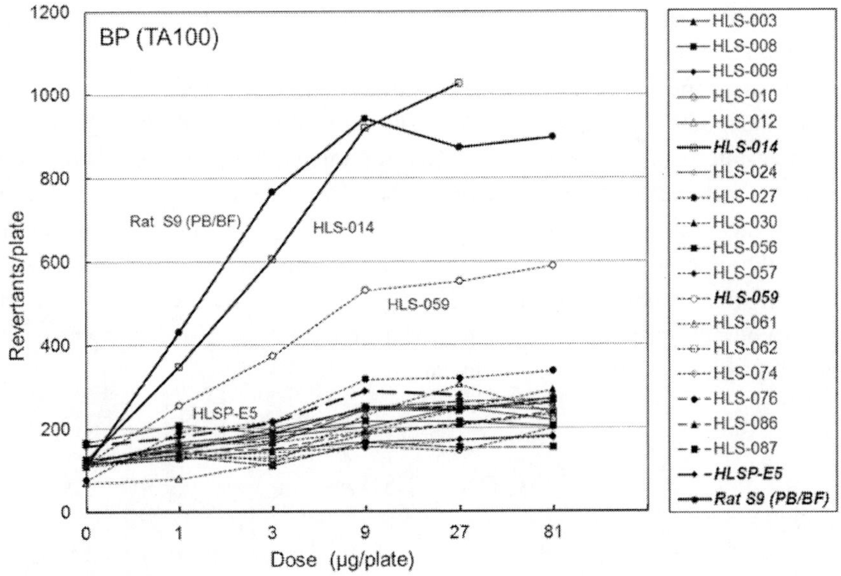

Data are taken from our study [149].

Figure 9. Mutagenicity of BP in Ames test using 18 individual human donor liver S9 fractions and S9 fraction prepared from pooled human donor livers (lot. HLSP-E5). Rat S9 (PB/BF) fraction was also used as a (positive) reference. Mutagenicity in the presence of S9 (1.1 mg protein/plate) with an NADPH-generating system was assayed towards the *S. typhimurium* TA100 test strain by the preincubation method at 37°C for 20 minutes. Human livers used were legally procured from NDRI (National Disease Research Interchange) in Philadelphia, USA, with permission to use only for research, based on the international partnership between NDRI and the Human and Animal Bridge Group (HAB) in Japan, and each human S9 fraction was prepared at the Biomedical Research Institute, Chiba, Japan.

Possible Environmental and Genetic Factors Increasing Susceptibility to BP-Induced CRC

Susceptibility to human CRC is greatly affected by other environmental and genetic factors, although it is not fully understood [1, 4, 8, 25, 104, 151]. For example, feeding Western-style diets to mice for 2 years led to the development of gross colonic lesions that were histologically classified as dysplastic crypts and focal hyperplasias with or without atypical nuclei [151]. Epidemiological studies have revealed that obesity, especially visceral adipose tissue, raises the risk of CRC, supported by animal experiments [152]. Ramesh and his colleagues showed the induction of colonic neoplasia in Apc^{Min} mice with a higher incidence of colon cancer development in the presence of dietary fat after exposure to BP [104]. Culp *et al.* [25] administered either BP or two coal tar mixtures in the diets of B6C3F1 mice for 2 years. BP alone only induced tumors of the alimentary tract, whereas the coal tar mixtures also induced liver and lung tumors, suggesting that coal tar components other than BP can exert a different tumorigenic action. Genetic polymorphisms in drug–metabolizing enzymes, DNA repair enzymes, or fatty acid metabolism have been suggested to play an important role in CRC susceptibility [37, 151, 153-158].

Increased risk of CRC is known to be found in patients with inflammatory bowel disease (IBD); the cumulative incidence of CRC in IBD patients ranged from 7.6% to 18.4% in a 30-year post-diagnosis [160, 161]. Intestinal microbiota play a key role in the pathogenesis of IBD, and this is supported by a colitis-associated colon cancer model using AOM-exposed, conventionalized-interleukin-10 knock-out ($Il10^{-/-}$) mice, showing that bacteria-mediated colitis enhances colorectal tumorigenesis [161]. Iron-deficiency anemia is a common complication of ulcerative colitis that results from chronic colorectal bleeding, and anemia is remedied in these cases by oral or parenteral iron supplementation. Dietary ion supplementation was shown to enhance DSS-induced colitis and associated with colorectal carcinoma development in mice [162]. Thus, iron therapy or iron overload carry a potential risk of enhancing colitis and inducing colon cancer [163].

Conclusion

BP has recently been classified into Group 1 (human carcinogen) by IARC [19]. The organs/tissues exposed by inhalation or skin contact, such as lung or skin, have drawn attention as target organs for BP carcinogenesis [1, 19]. On the contrary, the colon has not received so much attention because studies conducted so far have either failed to find or overlooked colonic tumors in normal animals [1, 19, 25, 26]. BP is also considered to be of little concern from a public health point of view according to the MOE approach applied for food [26]. In addition, epidemiological studies have provided controversial arguments on the contribution of BP to human CRC [1, 11, 154-156, 164-168]. One of the reasons may be the use of the analytical methods for the dection of BP-DNA adducts as a surrogate marker to evaluate BP-exposure level [169]. When there is no detection of DNA adducts, it does not necessarily mean that there is no exposure to chemicals of interest. This is because DNA adducts produced would be repaired, or diluted through DNA replication (cell divisions). Therefore, development or use of more appropriate biomonitoring methods will be needed.

BP itself is most unlikely to cause colonic cancer in normal animals, although available data are still very limited. However, it has become evident that BP can induce mutations in the murine colon at a high frequency by oral exposure [86, 88, 93, 107]. Thus, BP may lack the potency to promote/progress BP-initiated (mutated) colonic epithelial cells to colonic cancer. Another possible explanation for the insufficient colonic carcinogenicity of BP is that mutated colonic epithelial cells do not possess a sufficient property to work as cancer-initiating cells because they do not have such mutational alterations (mutational spectrum) that are capable of the activation or suppression of the related oncogenes or suppressor genes [4, 57, 58]. Moreover, there seems to be a large inter-individuality among humans with susceptibility to BP-induced mutagenicity, as suggested by the Ames test data using multiple human liver S9 fractions [149].

Increased risk of human CRC has been become widely acknowledged to be associated with IBD [2, 4, 159, 160]. Recently, we showed that BP can rapidly induce multiple colonic neoplasias in a colon colitis-associated cancer model using normal mice [133]. Increasing laboratory studies suggest the presence of other environmental and genetic factors that enhance or modulate the incidence of colonic tumors or the magnitude of colonic inflammation, such as dietary fats, obesity, or microbiota [1, 4, 104, 154-163]. Recent epidemiological studies suggest that abnormal epigenetic alterations are involved in human CRC [4, 72, 79, 138]. Figure 10 shows our working hypothesis for mechanisms of BP-induced colon carcinogenesis in the DSS-induced colon colitis model. Further studies on the carcinogenic risk of PAHs including BP, particularly in the colon, are strongly recommended.

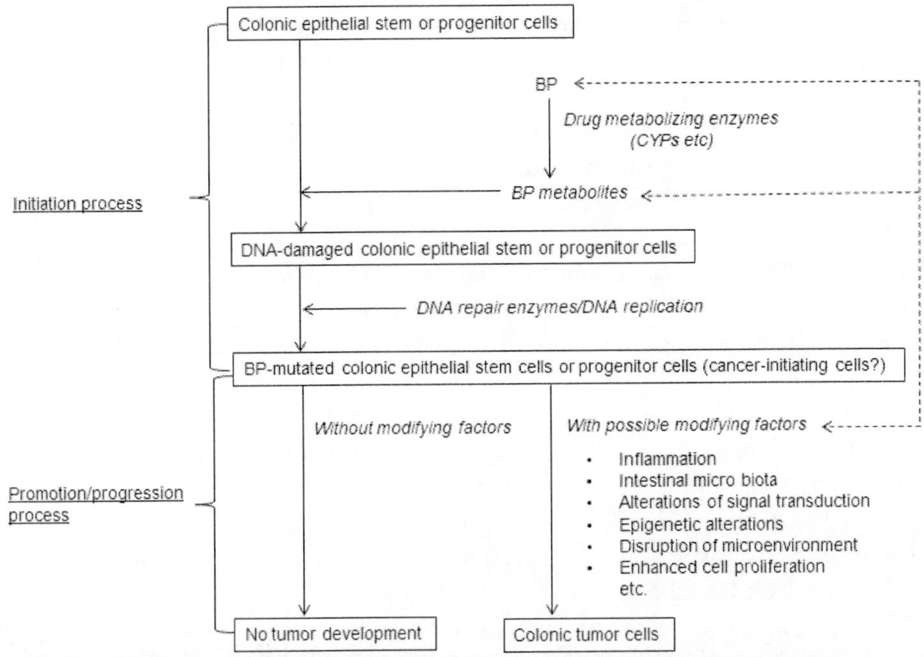

Figure 10. Our working hypothesis for mechanisms of BP-induced colon carcinogenesis in the presence of DSS, a colonic colitis-inducer.

ACKNOWLEDGMENTS

The authors wish to especially thank Drs. Kazuo Tsukidate and Satoru Hosokawa of Eisai Co., Ltd. for their valuable and continuous advice and encouragement.

REFERENCES

[1] Diggs DL, Huderson AC, Harris KL, Myers JN, Banks LD, Rekhadevi PV, Niaz MS, Ramesh A. Polycyclic aromatic hydrocarbons and digestive tract cancers: a perspective. *Journal of Environmental Science and Health, Part C.* 2011; 29: 324-357.

[2] Tanaka T. Colorectal carcinogenesis: Review of human and experimental animal studies, *J Carcinog*. 2009; 8: 1-19.

[3] National Cancer Institute-Surveillance Epidemiology and End Results (NCI-SEER). Cancer of colon and rectum. 2008. http://seer.cancer.gov/statfacts/html/colorect.html.

[4] Weinberg RA. The biology of cancer. New York: Garland Science; 2007.

[5] Ames BN, Magaw R, Gold LS. Ranking possible carcinogenic hazards. *Science* 1987; 236: 271-280.

[6] Lijinsky W. *N*-Nitroso compounds in the diet. *Mutat Res.* 1999; 443: 129-138.

[7] Woutersen RA, Appel MJ, van Garderen-Hoetmer A, Wijnands, MVW. Dietary fat and carcinogenesis. *Mutat Res.* 1999; 443: 111-127.

[8] Sugimura T. Nutrition and dietary carcinogens. *Carcinogenesis* 2000; 21; 387-395.

[9] de Kok TMCM, van Maanen JMS. Evaluation of fecal mutagenicity and colorectal cancer risk. *Mutat Res.* 2000; 3: 3-101.

[10] Bernstein H, Bernstein C, Payne CM, Dvorakova K, Garewal H. Bile acids as carcinogens in human gastrointestinal cancers. *Mutat Res.* 2005; 589: 47-65.

[11] Phillips DH. Polycyclic aromatic hydrocarbons in the diet. *Mutat Res.* 1999; 443: 139-147.

[12] Schoket B. DNA damage in humans exposed to environmental and dietary polycyclic aromatic hydrocarbons. *Mutat Res.* 1999; 424: 143-153.

[13] Phillips DH, Martin FL, Williams, JA, Wheat, LMC, Nolan L, Cole KJ, Grover PL. Mutagens in human breast lipid and milk: the search for environmental agents that initiate breast cancer. *Environ Mol Mutag.* 2002; 39: 143-149.

[14] Tajima K, Tominaga S. Dietary habits and gastro-intestinal cancers: A comparative case-control study of stomach and large intestinal cancers in Nagoya, Japan. *Jpn J Cancer Res (Gann),* 1985; 76: 705-716.

[15] Tominaga S, Kuroishi T. An ecological study on diet/nutrition and cancer in Japan. *Int J Cancer*, 1997; 10: S2-6.

[16] Kuriki K, Tajima K. The increasing incidence of colorectal cancer and the preventive strategy in Japan. *Asian Pac J Cancer Prev.* 2006; 7: 495-501.

[17] Bingham SA. Diet and colorectal cancer prevention. *Biochem Soc Trans.* 2000; 28: 12-16.

[18] Hutcheon DE, Kantrowitz J, Van Gelder RN, Flynn E. Factors affecting plasma benzo[a]pyrene levels in environmental studies. *Environ Res.* 1983; 32:104-110.

[19] IARC. Some non-heterocyclic polycyclic aromatic hydrocarbons and some related exposures. *IARC Monogr Eval Carcinog Risks Hum*. 2010; 92: 1–853.

[20] Devanesan PD, Higginbotham S, Ariese F, Jankowiak R, Suh M, Small GJ, Cavalieri EL, Rogan EG. Depurinating and stable benzo[*a*]pyrene-DNA adducts formed in isolated rat liver nuclei. *Chem Res Toxicol.* 1996; 9: 1113-1116.

[21] Flesher JW, Myers SR. Rules of molecular geometry for predicting carcinogenic activity of unsubstituted polynuclear aromatic hydrocarbons. *Teratog Carcinog Mutagen.* 1991; 11: 41-54.

[22] Conney AH. Induction of microsomal enzymes by foreign chemicals and carcinogenesis by polycyclic aromatic hydrocarbons: G. H. A. Clowes memorial lecture. *Cancer Res.* 1982; 42: 4875-4917.

[23] Gold LS, Slone TH, Manley NB, Bernstein L. Target organs in chronic bioassays of 533 chemical carcinogens. *Environ Health Perspect.* 1991; 93: 233-246.

[24] Hakura A, Shimada H, Nakajima M, Sui H, kitamoto S, Suzuki S, Satoh T. *Salmonella*/human S9 mutagenicity test: a collaborative study with 58 compounds. *Mutagenesis* 2005; 20: 217-228.

[25] Culp SJ, Gaylor DW, Sheldon WG, Goldstein LS, Beland FA. A comparison of the tumors induced by coal tar and benzo[*a*]pyrene in a 2-year bioassay. *Carcinogenesis* 1998; 19: 117-124.

[26] Benford D, DiNovi M, Setzer RW. Application of the margin-of-exposure (MoE) approach to substances in food that are genotoxic and carcinogenic e.g.: Benzo[*a*]pyrene and polycyclic aromatic hydrocarbons. *Food Chem Toxicol.* 2010; 48: S42-S48.

[27] van Maanen JMS, Moonen EJC, Maas LM, Kleinjans JCS, van Schooten FJ. Formation of aromatic DNA adducts in white blood cells in relation to urinary excretion of 1-hydroxypyrene during consumption of grilled meat. *Carcinogenesis* 1994; 15: 2263-2268.

[28] Alexandrov K, Rojas M, Kadlubar FF, Lang NP, Bartsch H. Evidence of *anti*-benzo[*a*]pyrene diolepoxide-DNA adduct formation in human colon mucosa. *Carcinogenesis* 1996; 17: 2081-2083.

[29] Shimada T, Okuda Y. Metabolic activation of environmental carcinogens and mutagens by human liver microsomes. *Biochem Pharmacol.* 1988; 37: 459-465.

[30] Shimada T, Martin MV, Pruess-Schwartz D, Marnett LJ, Guengerich, FP. Roles of individual human cytochrome P-450 enzymes in the bioactivation of benzo(a)pyrene, 7,8-dihydroxy-7,8-dihydrobenzo(a)pyrene, and other dihydrodiol derivatives of polycyclic aromatic hydrocarbons. *Cancer Res.* 1989; 49: 6304-6312.

[31] McManus ME, Burgess WM, Veronese ME, Huggett A, Quattrochi LC, Tukey RH. Metabolism of 2-acetylaminofluorene and benzo(a)pyrene and activation of food-derived heterocyclic amine mutagens by human cytochromes P-450. *Cancer Res.* 1990; 50: 3367-3376.

[32] Guengerich FP, Shimada T. Oxidation of toxic and carcinogenic chemicals by human cytochrome P-450 enzymes. *Chem Res Toxicol.* 1991; 4: 391-407.

[33] Shimada T, Yun CH, Yamazaki H, Gautier C, Beaune PH, Guengerich FP. Characterization of human lung microsomal cytochrome P450 1A1 and its role in the oxidation of chemical carcinogens. *Mol Pharmacol.* 1992; 41: 856-864.

[34] Shimada T, Hayes CL, Yamazaki H, Amin S, Hecht SS, Guengerich FP, Sutter TR. Activation of chemically diverse procarcinogens by human cytochrome P450 1B1. *Carcinogenesis* 1996; 56: 2979-2984.

[35] Gonzalez FJ, Kimura S. Study of P450 function using gene knockout and transgenic mice. *Arch Biochem Biophys.* 2003; 409: 153-158.

[36] Uno S, Dalton TP, Dragin N, Curran CP, Derkenne S, Miller ML, Shertzer HG, Gonzalez FJ, Nebert DW. Oral benzo[*a*]pyrene in *Cyp1* knockout mouse lines: CYP1A1 important in detoxification, CYP1B1 metabolism required for immune damage independent of total-body burden and clearance rate. *Mol Pharmacol.*2006; 69: 1103-1114.

[37] Shimada T. Xenobiotic-metabolizing enzymes involved in activation and detoxification of carcinogenic polycyclic aromatic hydrocarbons. *Drug Metab Pharmacokinet.* 2006; 21: 257-276.

[38] Robinson JR, Felton JS, Levitt RC, Thorgeirsson SS, Nebert DW. Relationship between "aromatic hydrocarbon responsiveness" and the survival times in mice treated with various drugs and environmental compounds. *Mol Pharmacol.* 1975; 11: 850-865.

[39] Sivarajah K, Lasker JM, Eling TE. Prostaglandin synthetase-dependent cooxidation of (±)-benzo[*a*]pyrene-7,8-dihydrodiol by human lung and other mammalian tissues. *Cancer Res.* 1981; 41: 1834-1839.

[40] Sivarajah K, Jones KG, Fouts JR, Devereux T, Shirley JE, Eling TE. Prostaglandin synthetase and cytochrome P-450-dependent metabolism of (±)benzo[*a*]pyrene-7,8-dihydrodiol by enriched populations of rat clara cells and alveolar type II cells. *Cancer Res.* 1983; 43: 2632-2636.

[41] Kozack RE, Loechler EL. Molecular modeling of the major adduct of (+)-*anti*-B[*a*]PDE(N^2-dG) in the eight conformations and the five DNA sequences most relevant to base substitution mutagenesis. *Carcinogenesis* 1999; 20: 85-94.

[42] Schurter EJ, Yeh HJ, Sayer JM, Lakshman MK, Yagi H, Jerina DM, Gorenstein DG. NMR solution structure of a nonanucleotide duplex with a dG mismatch opposite a 10R adduct derived from trans addition of a deoxyadenosine N6-amino group to (-)-(7S,8R,9R,10S)-7,8-dihydroxy-9,10-epoxy-7,8,9,10-tetrahydrobenzo[a]pyrene. *Biochemistry* 1995; 34: 1364-1375.

[43] Yeh HJ, Sayer JM, Liu X, Altieri AS, Byrd RA, Lakshman MK, Yagi H, Schurter EJ, Gorenstein DG, Jerina DM. NMR solution structure of a nonanucleotide duplex with a dG mismatch opposite a 10S adduct derived from trans addition of a deoxyadenosine N6-amino group to (+)-(7R,8S,9S,10R)-7,8-dihydroxy-9,10-epoxy-7,8,9,10-tetrahydrobenzo[a]pyrene: an unusual syn glycosidic torsion angle at the modified dA. *Biochemistry* 1995; 34: 13570-13581.

[44] Glick J, Xiong W, Lin Y, Noronha AM, Wilds, CJ, Vouros P. The influence of cytosine methylation on the chemoselectivity of benzo[a]pyrene diol epoxide-oligonucleotide adducts determined using nanoLC/MS/MS. *J Mass Spectrom.* 2009; 44: 1241-1248.

[45] Kulkarni MS, Anderson MW. Persistence of benzo(*a*)pyrene metabolite: DNA adducts in lung and liver of mice. *Cancer Res.* 1984; 44: 97-101.

[46] Stowers SJ, Anderson MW. Ubiquitous binding of benzo(*a*)pyrene metabolites to DNA and protein in tissues of the mouse and rabbit. *Chem Biol Interact.* 1984; 51: 151-166.

[47] Stowers SJ, Anderson MW. Formation and persistence of benzo(*a*)pyrene metabolite-DNA adducts. *Environ Health Perspect.* 1985; 62: 31-39.

[48] Helleberg H, Xu H, Ehrenberg L, Hemminki K, Rannung U, Törnqvist M. Studies of dose distribution, premutagenic events and mutation frequencies for benzo[*a*]pyrene aiming at low dose cancer risk estimation. *Mutagenesis* 2001; 16: 333-337.

[49] Yamada Y, Mori H. Multistep carcinogenesis of the colon in *Apc*$^{min/+}$ mouse. *Cancer Sci.* 2007; 98: 6-10.

[50] McCart AE, Vickaryous NK, Silver A. *Apc* mice: Models, modifiers and mutants. *Pathol Res Pract.* 2008; 204: 479-490.

[51] Sattar A, Hewer A, Phillips DH, Campbell FC. Metabolic proficiency and benzo[*a*]pyrene DNA adduct formation in Apc^{Min} mouse adenomas and uninvolved mucosa. *Carcinogenesis* 1999; 20: 1097-1101.

[52] Diggs DL, Harris KL, Rekhadevi PV, Ramesh A. Tumor microsomal metabolism of the food toxicant, benzo[*a*]pyrene, in Apc(Min) mouse model of colon cancer. *Tumour Biol* 2012; 33: 1255-1260.

[53] Schurdak ME, Randerath K. Effect of route of administration on tissue distribution of DNA adducts in mice: comparison of *7H*-dibenzo(c,g)carbazole, benzo(*a*)pyrene, and 2-acetylaminofluorene. *Cancer Res.*1989; 49: 2633-2638.

[54] Dunn BP. Wide-range linear dose-response curve for DNA binding of orally administered benzo(a)pyrene in mice. *Cancer Res*. 1983; 43: 2654-2658.

[55] Bi X, Slater DM, Ohmori H, Vaziri C. DNA polymerase κ is specifically required for recovery from the benzo[*a*]pyrene-dihydrodiol epoxide (BPDE)-induced S-phase checkpoint. *J Biol Chem* 2005; 280: 22343-22355.

[56] Rodriguez H, Loechler EL. Mutational specificity of the (+)-*anti*-diol epoxide of benzo[*a*]pyrene in a *supF* gene of an *Escherichia coli* plasmid: DNA sequence context influences hotspots, mutational specificity and the extent of SOS enhancement of mutagenesis. *Carcinogenesis* 1993; 14: 373-383.

[57] Mass MJ, Jeffers AJ, Ross JA, Nelson G, Galati AJ, Stoner GD, Nesnow S. Ki-ras oncogene mutations in tumors and DNA adducts formed by benz[j]aceanthrylene and benzo[a]pyrene in the lungs of strain A/J mice. *Mol Carcinog.* 1993; 8; 186-192.

[58] Chakravarti D, Venugopal D, Mailander PC, Meza JL, Higginbotham S, Cavalieri EL, Rogan EG. The role of polycyclic aromatic hydrocarbon-DNA adducts in inducing mutations in mouse skin. *Mutat Res.* 2008; 649: 161-178.

[59] Park JH, Mangal D, Frey AJ, Harvey RG, Blair IA, Penning TM. Aryl hydrocarbon receptor facilitates DNA strand breaks and 8-oxo-2'-deoxyguanonsine formation by the aldo-keto reductase product benzo[*a*]pyrene-7,8-dione. *J Biol Chem*. 2009; 284: 29725-29734.

[60] Balu N, Padgett WT, Lambert GR, Swank AE, Richard AM, Nesnow S. Identification and characterization of novel stable deoxyguanosine and deoxyadenosine adducts of benzo[*a*]pyrene-7,8-quinone from reactions at physiological pH. *Chem Res Toxicol.* 2004; 17: 827-838.

[61] Nesnow S, Nelson G, Padgett WT, George MH, Moore T, King LC, Adams LD, Ross JA. Lack of contribution of covalent benzo[*a*]pyrene-7,8-quinone-DNA adducts in benzo[*a*]pyrene-induced mouse lung tumorigenesis. *Chem Biol Interact* 2010; 186: 157-165.

[62] Kumar M, Sharma VL, Sehgal A, Jain M. Protective effects of green and white tea against benzo[*a*]pyrene induced oxidative stress and DNA damage in murine model. *Nutr Cancer* 2012; 64: 300-306.

[63] Long DJ, Waikel RI, Wang X-J, Perlaky L, Roop DR, Jaiswal AK. NAD(P)H: quinone oxidoreductase 1 deficiency increases susceptibility to benzo[*a*]pyrene-induced mouse skin carcinogenesis. *Cancer Res.* 2000; 60; 5913-5915.

[64] Cavalieri EL, Rogan EG. Central role of radical cations in metabolic activation of polycyclic aromatic hydrocarbons. *Xenobiotica* 1995; 25: 677-688.

[65] Melendez-Colon VJ, Luch A, Seidel A, Baird WM. Cancer initiation by polycyclic aromatic hydrocarbons result from formation of stable DNA adducts rather than apurinic sites. *Carcinogenesis* 1999; 20: 1885-1891.

[66] Flesher JW, Stansbury KH, Sydnor KL. *S*-Adenosyl-L-methionine is a carbon donor in the conversion of benzo[*a*]pyrene to 6-hydroxybenzo[*a*]pyrene by rat liver S-9, *Cancer Letters* 1982; 16: 91-94.

[67] Stansbury KH, Flesher JW, Gupta RC. Mechanism of aralkyl-DNA adduct formation from benzo[a]pyrene in vivo. *Chem Res Toxicol.* 1994; 7: 254-259.

[68] Shimizu Y, Nakatsuru Y, Ichinose M, Takahashi Y, Kume H, Mimura J, Fujii-Kuriyama Y, Ishikawa T. Benzo[*a*]pyrene carcinogenicity is lost in mice lacking the aryl hydrocarbon receptor. *Proc Natl Acad Sci USA* 2000; 97: 779-782.

[69] Tekpli X, Rissel M, Huc L, Catheline D, Sergent O, Rioux V, Legrand P, Holme JA, Dimanche-Boitrel M, Legadic-Gossmann D. Membrane remodeling, an early event in benzo[*a*]pyrene-induced apoptosis. *Toxicol Applied Pharmacol.* 2010; 243: 68-76.

[70] Chen S, Nguyen N, Tamura K, Karin M, Tukey RH. The role of the Ah receptor and p38 in benzo[*a*]pyrene-7,8-dihydrodiol and benzo[*a*]pyrene-7,8-dihydrodiol-9,10-epoxide-induced apoptosis. *J Biol Chem.* 2003; 278: 19526-19533.

[71] Nebert DW, Roe AL, Dieter MZ, Solis WA, Yang Y, Dalton TP. Role of the aromatic hydrocarbon receptor and [Ah] gene battery in the oxidative stress response, cell cycle control, and apoptosis. *Biochem Pharmacol* 2000; 59: 65-85.

[72] Sawan C, Vaissière T, Murr R, Herceg Z. Epigenetic drivers and genetic passengers on the road to cancer. *Mutat Res*. 2008; 642: 1-13.

[73] Sadikovie B, Rodenhiser DI. Benzopyrene exposure disrupts DNA methylation and growth dynamics in breast cancer cells. *Toxicol Applied Pharmacol.* 2006; 216: 458-468.

[74] Chen JX, Zheng Y, West M, Tang MS. Carcinogens preferentially bind at methylated CpG in the p53 mutational hot spots. *Cancer Res.* 1998; 58: 2070-2075.

[75] Yoon JH, Smith LE, Feng Z, Tang M, Lee CS, Pfeifer GP. Methylated CpG dinucleotides are the preferential targets for G-to-T transversion mutations induced by benzo[a]pyrene diol epoxide in mammalian cells: similarities with the p53 mutation spectrum in smoking-associated lung cancers, *Cancer Res*. 2001; 61: 7110-7117.

[76] Rogakou EP, Pilch DR, Orr AH, Ivanova VS, Bonner WM. DNA double-stranded breaks induce H2AX phosphorylation on serine 139. *J Biol Chem.* 1998; 273: 5858-5868.

[77] Srivastava N, Gochhait S, de Boer P, Bamezail RNK. Role of H2AX in DNA damage response and human cancers. *Mutat Res*. 2009; 681: 180-188.

[78] Fernandez-Capetillo O, Lee A, Nussenzweig M, Nussenzweig A. H2AX: the histone guardian of the genome. *DNA Repair* 2004; 3: 959-967.

[79] LeBaron MJ, Rasoulpour RJ, Klapacz J, Ellis-Hutchings RG, Hollnagel HM, Gollapudi BB. Epigenetics and chemical safety assessment. *Mutat Res*. 2010; 705: 83-95.

[80] Mattsson A, Jernström B, Cotgreave IA, Bajak E. H2AX phosphorylation in A549 cells induced by the bulky and stable DNA adducts of benzo[*a*]pyrene and dibenzo[*a,l*]pyrene diol epoxides. *Chem-Biol Interact.* 2009; 177: 40-47.

[81] Yan C, Wu W, Li H, Zhang G, Duerksen-Hughes PJ, Zhu X, Yang J. Benzo[a]pyrene treatment leads to changes in nuclear protein expression and alternative splicing. *Mutat Res.* 2010; 686: 47-56.

[82] Burchiel SW, Luster MI. Signalling by environmental polycyclic aromatic hydrocarbons in human lymphocytes. *Clin Immunol.* 2001; 98: 2-10.

[83] Allan LL, Schiezinger JJ, Shansab M, Sherr DH. CYP1A1 in polycyclic aromatic hydrocarbon-induced B lymphocyte growth suppression. *Biochem Biophys Res Commun.* 2006; 342: 227-235.

[84] Pratt I, Barlow S, Kleiner J, Larsen JC. The influence of thresholds on the risk assessment of carcinogens in food. *Mutat Res* 2009; 678: 113-117.

[85] Barlow S, Schlatter J. Risk assessment of carcinogens in food. Toxicol Applied Pharmacol. 2010; 243: 180-190.

[86] Hakura A, Tsutsui Y, Sonoda J, Kai J, Imade T, Shimada M, Sugihara Y, Mikami T. Comparison between *in vivo* mutagenicity and carcinogenicity in multiple organs by benzo[*a*]pyrene in the *lacZ* transgenic mouse (Muta™Mouse). *Mutat Res.* 1998; 398: 123-130.

[87] Hakura A, Sonoda J, Tsutsui Y, Mikami T, Imade T, Shimada M, Yaguchi S, Yamanaka M, Tomimatsu M, Tsukidate K. Toxicity profile of benzo[*a*]pyrene in the male *lacZ* transgenic mouse (MutaMouse) following oral administration for 5 consecutive days. *Regul Toxicol Pharmacol.* 1998; 27: 273-279.

[88] Hakura A, Tsutsui Y, Sonoda J, Mikami T, Tsukidate K, Sagami F, Kerns WD. Multiple organ mutation in the *lacZ* transgenic mouse (Muta™Mouse) 6 months after oral treatment (5 days) with benzo[*a*]pyrene. *Mutat Res.* 1999; 426: 71-77.

[89] Hakura A, Tsutsui Y, Sonoda J, Tsukidate K, Mikami T, Sagami F. Comparison of the mutational spectra of the *lacZ* transgene in four organs of the Muta™Mouse treated with benzo[*a*]pyrene: target organ specificity. *Mutat Res.* 2000; 447: 239-247.

[90] Gossen JA, de Leeuw WJF, Tan CHT, Lohman PHM, Berends F, Knook DL, Zwarthoff EC, Vijg J. Efficient mouse of integrated shuttle vectors from transgenic mice: a model for studying mutations in vivo. *Proc Natl Acad Sci USA.* 1989; 86: 7981-7985.

[91] Heddle JA, Swiger RR. Risk estimation from somatic mutation assays. *Mutat Res.* 1996; 365: 107-117.

[92] Gonzalez FJ. Use of transgenic animals in carcinogenesis studies. *Mol Carcinog.* 1996; 16: 63-67.

[93] Nohmi T, Suzuki T, Masumura K. Recent advances in the protocols of transgenic mouse mutation assays. *Mutat Res.* 2000; 455: 191-215.

[94] Lambert IB, Singer TM, Boucher SE, Douglas GR. Detailed review of transgenic rodent mutation assays. *Mutat Res.* 2005; 590: 1-280.

[95] Tsutsui Y, Tsukidate K, Hakura A. Sequence of the *lacZ* transgene of the Muta™Mouse. *Environ Mutagen Res.* 1999; 21: 39-43.

[96] Wang CX, Watanabe K, Weisburger JH, Williams GM. Induction of colon cancer in inbred Syrian Hamsters by intrarectal administration of benzo(a)pyrene, 3-methylchoranthrene and N-methyl-N-nitrosourea. *Cancer Lett.* 1985; 27: 309-314.

[97] Van Oostrom CT, Boeve M, van Den Berg J, de Vries A, Dollé ME, Beems RB, van Kreijl CF, Vijg J, van Steeg H. Effect of heterozygous loss of p53 on benzo[a]pyrene-induced mutations and tumors in DNA repair-deficient XPA mice. *Environ Mol Mutagen* 1999; 34: 124-130.

[98] Hoogervorst EM, de Vries A, Beems RB, van Oostrom CT, Wester PW, Vos JG, Bruins W, Roodbergen M, Cassee FR, Vijg J, van Schooten FJ, van Steeg H. Combined oral benzo[a]pyrene and inhalatory ozone exposure have no effect on lung tumor development in DNA repair-deficient Xpa mice. *Carcinogenesis* 2003; 24: 613-619.

[99] van Steeg H, Klein H, Beems RB, van Kreijl CF. Use of DNA repair-deficient XPA transgenic mice in short-term carcinogenicity testing. *Toxicol Pathol.* 1998; 26: 742-749.

[100] Ishikawa T, Ide F, Qin X, Zhang S, Takahashi Y, Sekiguchi M, Tanaka K, Nakatsuru Y. Importance of DNA repair in carcinogenesis: evidence from transgenic and gene targeting studies. *Mutat Res.* 2001; 477: 41-49.

[101] Tchou-Wong KM, Jiang Y, Yee H, LaRosa J, Lee TC, Pellicer A, Jagirdar J, Gordon T, Goldberg JD, Rom WN. Lung-specific expression of dominant-negative mutant p53 in transgenic mice increases spontaneous and benzo[*a*]pyrene-induced lung cancer. *Am J Respir Cell Mol Boil.* 2002; 27: 186-193.

[102] Kroese ED, Dortant PM, van Steeg H, van Oostrom CT, van der Houven van Oordt CW, van Kranen HJ, de Vries A, Wester PW, van Kreiji CF. Use of E mu-pim-1 transgenic mice short-term in vivo carcinogenicity testing: lymphoma induction by benzo[*a*]pyrene, but not by TPA, *Carcinogenesis* 1997; 18: 975-980.

[103] Lee BM, Kim HS, Kim IS, Ahn MY, Chun SA, Yoo SD, Park YA, Choi SM, Kim KB, Han SY, Park KL, Kwack SJ, Song HS, Lee KR, Bae HJ, Cho MH. Ras transgene expression in TG-AC mice treated with benzo[*a*]pyrene. *Biol Pharm Bull.* 2003; 26: 733-735.

[104] Harris DL, Washington MK, Hood DB, Roberts LJ 2nd, Ramesh A. Dietary fat-influenced development of colon neoplasia in *Apc*^{Min} mice exposed to benzo(a)pyrene. *Toxicol Pathol.* 2009; 37: 938-946.

[105] Halberg RB, Larsen MC, Elmergreen TL, Ko AY, Irving AA, Clipson L, Jefcoate CR. Cyp1b1 exerts opposing effects on intestinal tumorigenesis via exogenous and endogenous substrates. *Cancer Res.* 2008; 68: 7394-7402.

[106] Shimada H, Suzuki H, Itoh S, Hattori C, Matsuura Y, Tada S, Watanabe C. The micronucleus test of benzo[*a*]pyrene with mouse and rat peripheral blood reticulocytes. *Mutat Res.* 1992; 278: 165-168.

[107] Kosinska W, Pressentin MdMv, Guttenplan JB. Mutagenesis induced by benzo[*a*]pyrene in *lacZ* mouse mammary and oral tissues: comparisons with mutagenesis in other organs and relationships to previous carcinogenicity assays. *Carcinogenesis* 1999; 20: 1103-1106.

[108] Vries Ad, Dollê ET, Broekhof JLM, Muller JJA, Hroese ED, Kreijl CF, Capel PJA, Vijg J, Sttg Hv. Induction of DNA adducts and mutations in spleen, liver and lung of *XPA*-deficient/*lacZ* transgenic mice after oral treatment with benzo[*a*]pyrene: correlation with tumor development. *Carcinogenesis* 1997; 18: 2327-2332.

[109] Winton DJ, Blount MA, Ponder AJ. A clonal marker induced by mutation in mouse intestinal epithelium. *Nature* 1988; 333: 463-466.

[110] Park H-S, Goodlad RA, Wright NA. Crypt fission in the small intestine and colon: a mechanism for the emergence of G6PD locus-mutated crypts after treatment with mutagens. *Am J Pathol.* 1995; 147: 1416-1427.

[111] Kuraguchi M, Thomas GA, Williams ED. Somatic mutation of the glucose-6-phosphate dehydrogenase (g6pd) gene in colonic stem cells and crypt restricted loss of G6PD activity. *Mutat Res.* 1997; 379: 69-75.

[112] Yamada K, Suzuki T, Kohara A, Hayashi M, Hakura A, Mizutani T, Saeki K. Effect of 10-aza-substitution on benzo[*a*]pyrene mutagenicity in vivo and in vitro. *Mutat Res.* 2002; 521: 187-200.

[113] Hashimoto AH, Amanuma K, Hiyoshi K, Takano H, Masumura K, Nohmi T, Aoki Y. *In vivo* mutagenesis induced by benzo[*a*]pyrene instilled into the lung of *gpt* delta transgenic mice. 2005; 45: 365-373.

[114] Robles AI, Linke SP, Harris CC. The p53 network in lung carcinogenesis. *Oncogene* 2002; 21: 6898-6907.

[115] Cunningham ML. Matthews HB. Cell proliferation as a determining factor for the carcinogenicity of chemicals: Studies with mutagenic carcinogens and mutagenic noncarcinogens. *Toxicol Lett.* 1995; 82/83: 9-14.

[116] Wong WM, Wright NA. Cell proliferation in gastrointestinal mucosa. *J Clin Pathol.* 1999; 52: 321-333.

[117] Cohen SM, Arnold LL. Cell proliferation and carcinogenesis. *J Toxicol Pathol.* 2008; 21: 1-7.

[118] Kim H, Berstad A. Experimental colitis in animal models. *Scand J Gastroenterol.* 1992; 27: 529-537.

[119] Clapper ML, Cooper HS, Chang W-CL. Dextran sulfate sodium-induced colitis-associated neoplasia: a promising model for the development of chemopreventive interventions. *Acta Pharmacol Sin.* 2007; 28: 1450-1459.

[120] Rosenberg D, Giardina C, Tanaka T. Mouse models for the study of colon carcinogenesis. *Carcinogenesis* 2009; 30: 183-196.

[121] Westbrook AM, Szakmary A, Schiestl RH. Mechanisms of intestinal inflammation and development of associated cancers: lessons learned from mouse models. *Mutat Res.* 2010; 705: 40-59.

[122] Cooper HS, Murthy S, Kido K, Yoshitake H, Flanigan A. Dysplasia and cancer in the dextran sulfate sodium mouse colitis model. Relevance to colitis-associated neoplasia in the human: a study of histopathology, B-catenin and p53 expression and the role of inflammation. *Carcinogenesis* 2000; 21: 757-768.

[123] Okayasu I, Yamada M, Mikami T, Yoshida T, Kanno J, Ohkusa T. Dysplasia and carcinoma development in experimental colitis: Dysplasia and carcinoma development in a repeated dextran sulfate sodium-induced colitis model. *J Gastroenterol Hepatol.* 2002; 17: 1078-1083.

[124] Chang W-CL, Coudry RA, Clapper ML, Zhang X, Williams K-L, Spittle CS, Li T, Cooper HS. Loss of p53 enhances the induction of colitis-associated neolasia by dextran sulfate sodium. *Carcinogenesis* 2007; 28: 2375-2381.

[125] Tanaka T, Kohno H, Suzuki R, Yamada Y, Sugie S, Mori H. A novel inflammation-related mouse colon carcinogenesis model induced by azoxymethane and dextran sodium sulfate. *Cancer Sci.* 2003; 94: 965-973.

[126] Suzuki R, Kohno H, Sugie S, Tanaka T. Sequential observations on the occurrence of preneoplastic and neoplastic lesions in mouse colon treated with azoxymethane and dextran sodium sulfate. *Cancer Sci.*2004; 95: 721-727.

[127] Suzuki R, Kohno H, Sugie S, Nakagama H, Tanaka T. Strain differences in the susceptibility to azoxymethane and dextran sodium sulfate-induced colon carcinogenesis in mice. *Carcinogenesis* 2006; 27: 162-169.

[128] Onose J-I, Imai T, Hasumura M, Ueda M, Hirose M. Rapid induction of colorectal tumors in rats initiated with 1,2-dimethylhydrazine followed by dextran sodium sulfate treatment. *Cancer Lett.* 2003; 198: 145-152.

[129] Wang JG, Wang DF, Lv BJ, Si JM. A novel mouse model for colitis-associated colon carcinogenesis induced by 1,2-dimethylhydrazine and dextran sulfate sodium. *World J Gastroenterol.* 2004; 10: 2958-2962.

[130] Imai T, Fukuta K, Hasumura M, Cho Y-M, Ota Y, Takami S, Nakagama H, Hirose M. Significance of inflammation-associated regenerative mucosa characterized by Paneth cell metaplasia and β-catenin accumulation for the onset of colorectal carcinogenesis in rats treated with 1,2-dimethylhydrazine. *Carcinogenesis* 2007; 28: 2199-2206.

[131] Tanaka T, Suzuki R, Kohno H, Sugie S, Takahashi M, Wakabayashi K. Colonic adenocarcinomas rapidly induced by the combined treatment with 2-amino-1-methyl-6-phenylimidazo[4,5-*b*]pyridine and dextran sodium sulfate in male ICR mice possess β-*catenin* gene mutations and increases immunoreactivity for β-*catenin*, cyclooxygenase-2 and inducible nitric oxide synthase. *Carcinogenesis* 2005; 26: 229-238

[132] Nishikawa A, Imazawa T, Kuroiwa Y, Kitamura Y, Kanki K, Ishii Y, Umemura T, Hirose M. Induction of colon tumors in C57BL/6J mice fed MeIQx, IQ, or PhIP followed by dextran sulfate sodium treatment. *Toxicol Sci.* 2005; 84: 243-248.

[133] Hakura A, Seki Y, Sonoda J, Hosokawa S, Aoki T, Suganuma A, Kerns WD, Tsukidate K. Rapid induction of colonic adenocarcinoma in mice exposed to benzo[*a*]pyrene. *Food Chemical Toxicol.* 2011; 49: 2997–3001.

[134] Brittan M, Wright NA. Stem cell in gastrointestinal structure and neoplastic development. *Gut* 2004; 53: 899-910.

[135] Humphries A, Wright NA. Colonic crypt organization and tumorigenesis. *Nature Reviews* 2008; 8: 415-424

[136] Fukata M, Chen A, Vamadevan AS, Cohen J, Breglio K, Krishnareddy S, Hsu D, Xu R, Harpaz N, Dannenberg AJ, Subbaramaiah K, Cooper HS, Itzkowitz SH, Abreu MT. Toll-like receptor-4 promotes the development of colitis-associated colorectal tumors. *Gastroenterology* 2007; 133: 1869-1881.

[137] Fujisawa T, Nakajima A, Fujisawa N, Takahashi H, Ikeda I, Tomimoto A, Yonemitsu K, Nakajima N, Kudo C, Wada K, Kubota N, Terauchi Y, Kadowaki T, Nakagama H, Blumberg RS. Peroxisome proliferator-activated receptor γ (PPARγ) suppresses colonic epithelial cell turnover and colon carcinogenesis through inhibition of the β-catenin/T cell factor (TCF) pathway. *J Pharmacol Sci.* 2008; 106: 627-638.

[138] Barnett MB, Bermingham E, McNabb W, Bassett S, Armstrong K, Rounce J, Roy N. Investigating micronutrients and epigenetic mechanisms in relation to inflammatory bowel disease. *Mutat. Res.* 2010; 690: 71-80.

[139] Kundu JK, Surh Y. 1nflammation: Gearing the journey to cancer. *Mutat Res.* 2008; 659: 15-30.

[140] Grivennikov S, Karin E, Terzic J, Mucida D, Yu G-Y, Vallabhapurapu S, Scheller J, Rose-John S, Cheroutre H, Eckmann L, Karin M. IL-6 and Stat3 are required for survival of intestinal epithelial cells and development of colitis-associated cancer, *Cancer cell* 2009; 15: 103-113.

[141] Maron DM, Ames BN. Revised methods for the Salmonella mutagenicity test. *Mutat Res* 1983; 113: 173-215.

[142] Mortelmans K, Zeiger E. The Ames *Salmonella*/microsome mutagenicity test. *Mutat Res*. 2000; 455: 29-60.

[143] Beaune P, Lemestré-Cornet R, Kremers P, Albert A, Gielen J. The Salmonella/mammalian microsome mutagenicity test: comparison of human and rat livers as activating systems. *Mutat Res.* 1985; 156: 139-146.

[144] Le J, Jung R, Kramer M. Dedicated to Prof. Med. Helmut Kewitz in honour of his 65th birthday: effects of using liver fractions from different mammals, including man, on

results of mutagenicity assays in Salmonella typhimurium. *Food Chem Toxicol.* 1985; 23: 695-700.

[145] Neis JM, Yap SH, van Gemert PJL, Roelofs HMJ, Bos RP, Henderson PTh. Activation of mutagens by hepatocytes and liver 9000 x *g* supernatant from human origin in the *Salmonella typhimurium* mutagenicity assay: Comparison with rat liver preparations. *Mutat Res.* 1986; 164: 41-51.

[146] Phillipson CE, Ioannides C. Metabolic activation of polycyclic aromatic hydrocarbons to mutagens in the Ames test by various animal species including man. *Mutat Res* 1989; 211: 147-151.

[147] Johnson TE, Umbenhauer DR, Galloway SM. Human liver S-9 metabolic activation: proficiency in cytogenetic assays and comparison with phenobarbital/β-naphthoflavone or aroclor 1254 induced rat S-9. *Environ Mol Mutagen.* 1996; 28: 51-59.

[148] Hakura A, Suzuki S, Satoh T. Advantage of the use of human liver S9 in the Ames test. *Mutat Res.* 1999; 438: 29-36.

[149] Hakura A, Suzuki S, Sawada S, Sugihara T, Hori Y, Uchida K, Kerns WD, Sagami F, Motooka S, Satoh T. Use of human liver S9 in the Ames test: assay of three procarcinogens using human S9 derived from multiple donors. *Regul Toxicol Pharmacol.* 2003: 37: 20-27.

[150] Autrup H, Grafstrom RC, Brugh M, Lechner JF, Haugen A, Trump BF, Harris CC. Comparison of benzo(*a*)pyrene metabolism in bronchus, esophagus, colon, and duodenum from the same individual. *Cancer Res.* 1982; 42: 934-938.

[151] Risio M, Lipkin M, Newmark H, Yang K, Rossini FP, Steele VE, Boone CW, Kelloff GJ. Apoptosis, cell replication, and Western-style diet-induced tumorigenesis in mouse colon. *Cancer Res.* 1996; 56: 4910-4916.

[152] Frezza EE, Wachtel MS, Chiriva-Internati M. Influence of obesity on the risk of developing colon cancer. *Gut* 2006; 55: 285-291.

[153] Marchand LL, Hankin JH, Pierce LM, Sinha R, Nerurkar PV, Franke AA, Wilkens LR, Kolonel LN, Donlon T, Seifried A, Custer LJ, Lum-Jones A, Chang W. Well-done red meat, metabolic phenotypes and colorectal cancer in Hawaii. *Mutat Res.* 2002; 506-507: 205-214.

[154] Autrup H. Genetic polymorphisms in human xenobiotica metabolizing enzymes as susceptibility factors in toxic response. *Mutat Res.* 2000; 464: 65-76.

[155] Pavanello S, Clonfero E. Biological indicators of genotoxic risk and metabolic polymorphisms. *Mutat Res.* 2000; 463: 285-308.

[156] Girard H, Butler LM, Villeneuve L, Millikan RC, Sinha R, Sandler RS, Guillemette C. UGT1A1 and UGT1A9 functional variants, meat intake, and colon cancer among Caucasians and African-Americans. *Mutat Res.* 2008; 644: 56-63.

[157] Tomlinson IPM, Houlston RS, Montgomery GW, Sieber OM, Dunlop MG. Investigation of the effects of DNA repair gene polymorphisms on the risk of colorectal cancer. *Mutagenesis* 2012; 27: 219-223.

[158] Crous-Bou M, Rennert G, Salazar R, Rodriguez-Moranta F, Rennert HS, Lejbkowicz F, Kopelovich L, Lipkin SM, Gruber SB, Moreno V. Genetic polymorphisms in fatty acid metabolism genes and colorectal cancer. *Mutagenesis* 2012; 27: 169-176.

[159] Jess T, Loftus EV, Velayos FS, Harmsen WS, Zinsmeister AR, Smyrk TC, Schleck CD, Tremaine WJ, Melton LJ, Munkholm P, Sandborn WJ. Risk of intestinal cancer in

inflammatory bowel disease: a population-based study from olmsted country, Minnesota. *Gastroenterology* 2006; 130: 1039-1046.

[160] Rutter MD, Saunders BP, Wilkinson KH, Rumbles S, Schofield G, Kamm MA, Williams CB, Price AB, Talbot IC, Forbes A. Thirty-year analysis of a colonoscopic surveillance program for neoplasia in ulcerative colitis. *Gastroenterology* 2006; 130: 1030-1038.

[161] Uronis JM, Mühlbauer M, Herfarth HH, Rubinas TC, Jones GS, Christian J. Modulation of the intestinal microbiota alters colitis-associated colorectal cancer susceptibility. *PLoS ONE* 2009; 4: 1-9.

[162] Seril DN, Liao J, Ho K-LK, Warsi A, Yang CS, Yang G-Y. Dietary ion supplementation enhances DSS-induced colitis and associated colorectal carcinoma development in mice. *Dig Dis Sci.* 2002; 47: 1266-1278.

[163] Reifen R, Matas Z, Zeidel L, Berkovitch Z, Bujanover Y. Ion supplementation may aggravate inflammatory status of colitis in a rat model. *Dig Dis Sci.* 2000; 45: 394-397.

[164] Butler LM, Sinha R, Millikan RC, Martin CF, Newman B, Gammon MD, Ammerman AS, Sandler RS. Heterocyclic amines, meat intake, and association with colon cancer in a population-based study. *Am J Epidemiol.* 2003; 157: 434-445.

[165] Sinha R, Peters U, Cross AJ, Kulldorf M, Weissfeld JL, Pinsky PF, Rothman N, Hayes RB. Prostate, lung, colorectal, and ovarian cancer project team. Meat, meat cooking methods and preservation, and risk for colorectal adenoma. *Cancer Res* 2005; 65: 8034-8041.

[166] Lee BM, Shim GA. Dietary exposure estimation of benzo[*a*]pyrene and cancer risk assessment. *J Toxicol Environ Health A.* 2007; 70: 1391-1394.

[167] Tabatabaei SM, Heyworth JS, Knuiman MW, Fritschi L. Dietary benzo[*a*]pyrene intake from meat and the risk of colorectal cancer. *Cancer Epidemiol Biomarkers Prev.* 2010; 19: 3182-3184.

[168] Willet WC, Stampfer MJ, Colditz GA, Rosner BA, Speizer FE. Relation of meat, fat, and fiber intake to the risk of colon cancer in a prospective study among women. *N Engl J Med.* 1990; 323: 1664-1672.

[169] Boysen G, Hecht SS. Analysis of DNA and protein adducts of benzo[*a*]pyrene in human tissues using structure-specific methods. *Mutat Res.* 2003; 543: 17-30.

In: Pyrene
Editors: Petr Ruzicka and Tomas Kral

ISBN: 978-1-62417-291-5

Chapter 3

BACTERIAL METABOLISM OF PYRENE

O. S. Obayori*[*,1], M. O. Ilori[2] and O. O. Amund[2]
[1]Department of Microbiology, Lagos State University,
Ojo, Lagos, Nigeria
[2]Department of Microbiology, University of Lagos,
Akoka, Lagos, Nigeria

ABSTRACT

Pyrene is a peri-condensed four-ring polycyclic aromatic hydrocarbon (PAH). It is highly recalcitrant and a model compound for the study of degradative pathways of other PAHs such as benzo(a)pyrene. *Mycobacterium* species and some other actinomycetes were the first pyrene metabolizing bacteria to be described starting from 1988. Today, pyrene is known to be metabolized by a broad range of bacteria including Gram-negatives such as *Pseudomonas, Sphingomonas, Stenotromorpha, Alcaligenes, Achromobacter, Leclercia, Cycloclasticus and Burkholderia* among others. The main pathway of pyrene metabolism is via dioxygenation at the 4,5-position followed by *ortho* cleavage to produce 4,5-dihydroxy-4,5-dihydro phenanthrene and subsequent sequential cleavage leading to the tricarboxylic acid (TCA) cycle via O-phthalate. Another pathway, less common, proceeds through the cleavage of the two central rings at the 4, 5-and 9, 10-positions to produce a unique metabolite 6,6'-dihydroxy-2,2'-biphenyl dicarboxylic acid which also proceeds through O-phthalate but in fewer steps. Dioxygenation at the 1,2 position often leads to methylated dead-end products or detoxification. There are also reports of cometabolism of pyrene in the presence of co-substrates such as phenanthrene and fluoranthene. Bacterial metabolism of pyrene is modulated by factors such as pH, temperature, salinity, biosurfactants and humic acids amongst others. Products of incomplete metabolism of pyrene have been shown to inhibit degradation of other PAHs in the environment. In *Mycobacterium* species, especially, genes for pyrene metabolism are localized both on chromosomes and on plasmids. However, not much is known yet about the metabolism and genomics of Gram-negative pyrene degraders. Therefore, efforts are expected to be directed in this direction in the near future.

* Corresponding Author: Oluwafemi S. Obayori, Department of Microbiology, Faculty of Science, Lagos State University, Lagos, Nigeria, Phone: (234) 802-319-2652, E-mail: femiobayori@yahoo.com.

PYRENE: GENERAL DESCRIPTION

Pyrene is a high molecular weight peri-condensed four-ring polycyclic aromatic hydrocarbon (PAH) (Figure 1). It has a molecular weight of 202.3, vapour pressure of 0.0006 and solubility 0.135 mg/l (US Dept. of Health and Human Services, 1990). Pure pyrene is a colourless crystalline solid at ambient temperature; the presence of tetracene, a common contaminant, gives it a yellow colour. Pyrene can be derived from coal tar.

It is a regulated contaminant at sites polluted with petroleum (Kazunga and Aitken, 2000). Its accumulation in urban/regional areas due to anthropogenic generation is always of serious health concern. It is used as a model compound to study the biodegradation of high molecular weight PAHs because it is structurally similar to several carcinogenic PAHs (Kanaly and Harayama, 2000).

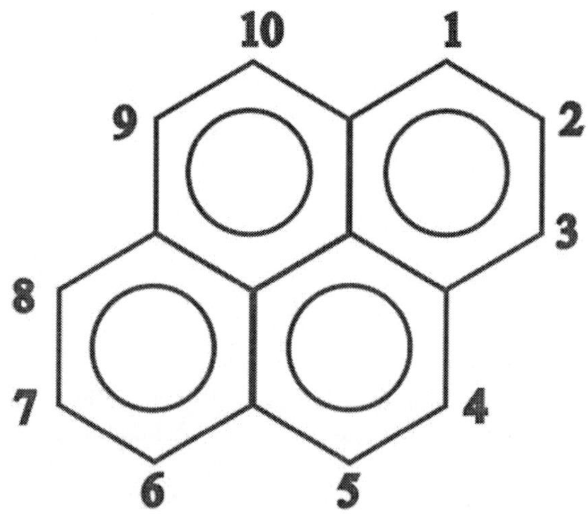

Figure 1. Molecular structure of pyrene.

SOLUBILITY AND REACTIVITY OF PYRENE

PAHs are hydrophobic compounds and their persistence within ecosystems is due in part to their low water solubility. This is because PAHs are composed of fused rings whose biochemical persistence arises from dense clouds of π-electrons on both sides of the ring structures, making them resistant to nucleophilic attack (Johnsen *et al.*, 2005). There is a positive correlation between increase in size of a PAH molecule, and its angularity, and its hydrophobicity and electrochemical stability. Pyrene has an aqueous solubility of 0.135 mg/l. It is more soluble than anthracene (0.015 mg/l) in spite of its higher molecular weight, but less soluble than naphthalene (30 mg/l) and phenanthrene (1-2 mg/l) (Table 1).

It is more recalcitrant than phenanthrene but less than benzo(a)anthracene and benzo(a)pyrene (Cerniglia, 1992). It is soluble in acetone, hexadecane, and ethyl acetate. Poor solubility in water means that pyrene is not readily available to biological systems since biochemical reactions take place in the aqueous phase – its bioavailability is low and thus it is highly stable in the environment and highly recalcitrant.

Table 1. Comparison of properties of pyrene and some other common PAHs

PAH	Mass Da	Vapour pressure (Pa)	Log K_{ow}	Solubility	Carcinogenicity
Pyrene	202.26	0.0006	5.88-6.7	0.12-0.18	-
Naphthalene	128.18	12.0	3.58	30	-
Acenaphthene	154.20	4.02	3.92	3.6	-
Acenaphthylene	155.20	3.87	3.90	3.88	-
Fluorene	166.23	0.13	4.18	2.00	-
Phenanthrene	178.24	0.0161	4.46-4.63	1-2	-
Anthracene	178.24	0.001	4.45	0.015	-
Fluoranthene	202.26	0.001	5.22	0.25	-
Benz[a]anthracene	228.30	$2.0x10^{-5}$	5.99	0.01	+
Chrysene	228.30	$6.08x10^{-7}$	5.01-7.10	0.0015-0.004	+
Benzo[a]pyrene	252.32	$7.0x10^{-7}$	5.78-6.5	0.001-0.006	+

K_{ow} – octanol-water coefficient.
Source: Cerniglia and Shuttleworth, 2002.

AROMATICITY OF PYRENE

Aromaticity can be defined as a chemical property in which a conjugated ring of unsaturated bonds, lone pairs, or empty orbitals exhibits a stabilisation stronger than would be expected by the stabilisation of conjugation alone. It can also be considered a manifestation of cyclic delocalization and of resonance (Balaban *et al.*, 2005). This is because electrons are free to cycle around circular arrangements of atoms, which are alternately single- and double-bonded to one another.

Thus, contrary to the hypothetical structure with dumb bell orbital clouds, the carbon atoms are sandwiched between clouds of л electrons. In reality all the electrons are evenly distributed around the atoms, which make them equal partners; they are coplanar.

The consequence of this is that while there are not enough electrons to form a double bond between any pair of atoms, there are "extra" electrons to strengthen all of the bonds on the ring equally. This is the orbital secret of enhanced chemical stability, compared to similar non-aromatic molecules. Aromatic compounds react by substitution rather than addition.

Although all PAHs are aromatic, the degree of aromaticity varies among the different types, just as the different rings in a molecule have different aromaticity. The aromaticity of pyrene ring is localized in specific regions of the molecular framework (Garcia-Cruz and Martinez-Magadan, 2007).

This is of great consequence for reactivity. The two central rings with the least aromaticity are more reactive, unlike in the case of anthracene where the aromaticity spreads out and is not restricted to particular ring or phenanthrene where there are two possible resonance structures. Furthermore, the clustering of rings and lack of angular structure, as in phenanthrene and some other PAHs, limits the variability of initial attack during its metabolism as we shall see shortly.

PYRENE DEGRADING BACTERIA

The first pyrene degrading bacterium to be isolated was *Mycobacterium* sp. Strain PYR1 (Heitkamp *et al.*, 1988a) later renamed correctly as *Mycobacterium vanbaalenii* strain PYR1 by Khan *et al.* (2002). It produces yellow pigmented scotochromogenic colonies with smooth convex surface. G+C content of DNA 66.6%. It is Gram positive, strongly acid fast rod. It metabolized 48% of pyrene supplied (50 ppm) within 72 hrs. It also mineralized naphthalene, phenanthrene, fluoranthrene, 3-methylcholanthrene, 1-nitropyrene and 6-nitrochrysene.

In the course of the next decade or so most of the strains found to utilize pyrene as sole carbon sources are mycobacteria and other norcadioforme actinomycetes (Dean-Ross and Cerniglia, 1996; Bastiaens *et al.*, 2000; Cheung and Kinkle, 2001). These organisms exhibit a kind of versatility not common with other groups known to metabolize PAHs (Heitkamp *et al.* 1988a; Grosser *et al.*, 1991; Walter *et al.*, 1991; Boldrin *et al.*, 1993; Kastner *et al.*, 1994; Schneider *et al.*, 1996; Churchill *et al.*, 1999; Kazunga and Atkiens 2000, Vila *et al.*, 2001). For instance, *M. vanbaaleni* PYR 1 has been shown to metabolise fluoranthene, phenanthrene, anthracene, 7, 12-dimethylbenzo[a]pyrene, benzo [a] pyrene as well as benzo[a]anthracene (Kelley *et al.*, 1993; Moody *et al.*, 2001, 2003, 2004, 2005).

Bastiaens *et al.* (2000) noted that specific physiological properties of the microorganisms involved in biodegradation of a hydrophobic substance might enhance the availability of the compound. The authors identified three mechanisms that promote the transfer of hydrophobic substrates, namely:

1. Production of biosurfactants or the production of specific cell surface components with emulsifying properties.
2. Uptake systems with high substrate affinity, which efficiently reduce concentration of substrate close to the cell surface thereby increasing the diffusive substrate flux.
3. Reduction of the distance between cells and substrate by means of cell surface structure which promote adhesion to hydrophobic surfaces.

According to these authors, the genera *Sphingomonas* and *Mycobacterium* seemed to be specialized in degrading less bioavailable compounds because of their peculiar outer cell wall layers; sphingolipid in *Sphingomonas* and mycolic acid in mycobacteria.

It has also been shown that mycobacteria present in the contaminated environment capable of utilizing high molecular weight PAHs such as pyrene far outstrip the number so far isolated (Cheung and Kinkle, 2001). The stringency of traditional enrichment techniques, dilution and over-growing of such slow-growing strains by fast-growing non-actinomycetes bacteria are some of the reasons that have been cited for the disparity between pristine condition, mycobacteria diversity and the range isolated in pure culture (Stach and Burns, 2002). Furthermore, it has been suggested that culture methods restrict the selection of bacteria to members that are competitive under the conditions of enrichment and are likely to represent a small fraction of those present in the original sample.

This notwithstanding, in the last decade or so, isolates spanning a wide range of actinomycetes, low G+C Gram-positives and Gram-negatives have been reported, indicating that bacteria capable of utilizing pyrene as sole source of carbon and energy are more diverse than thought hitherto. For instance, Hilyard *et al.* (2008) found seven pyrene degraders isolated from river sediments to be closely related to *Pusillimonas* sp., *Alcaligenes* sp., *Mycobacterium* sp., *Pseudomonas* sp., *Parvibaculum* sp., *Stappia* sp., and *Microbacterium* sp. Pyrene degraders in the literature are found among the following genera: *Bacillus, Corynebacterium, Micrococcus, Pseudomonas, Sphingomonas, Stenotromorpha, Alcaligenes, Achromobacter, Leclercia, Cycloclasticus and Burkholderia* and *Proteus* (Kanaly and Harayama, 2010). A comprehensive list of the best reported pyrene degraders in the last 24 years is presented in Table 2.

Sarma *et al.* (2004) isolated a newly discovered enteric bacterium *Leclercia adecarboxylata* PS4040 from subsurface soil collected from an oily-sludge storage pit at the Dugboi Oil field, India.

Table 2. Some of the best reported pyrene degraders

Isolate References	
Mycobacterium sp. PYRI	Heitkamp *et al.*, 1988
Mycobacterium sp. RJG II 135	Grosser *et al.*, 1991
Rhodococcus sp. UW1	Walter *et al.*, 1991
Mycobacterium sp. BB1	Boldrin *et al.*, 1993
Mycobacterium sp. VF1	Kastner *et al.*, 1994
Gordona sp. BP9	Kastner *et al.*, 1994
Mycobacterium sp.	Jimenez and Bartha, 1996
Mycobacterium flavescens	Dean-Ross and Cerniglia, 1996
Burkolderia cepacia	Juhasz *et al.*, 1997
Mycobacterium CHI	Churchill *et al.*, 1999
Mycobacterium sp. MR-1	Molina *et al.*, 1999
Mycobacterium gilvum LB208	Bastiaens *et al.*, 2000
Mycobacterium fredercksbergense FAn9^{T}	Williumsen *et al.*, 2001
Mycobacterium sp. API	Vila *et al.*, 2001
Mycobacterium sp. 6PY1	Krivobok *et al.*, 2003
Saccharotrix xinjiangensis . PYX-6	Hu *et al.*, 2003, Hu *et al.*2004
Mycobacterium pyrenivorans	Derz *et al.*, 2004
Mycobacterium sp. KMS	Miller *et al.*, 2004
Mycobacterium sp. MHP-1	Habe *et al.*, 2004
Leclercia adecarboxylata	Sarma *et al.*, 2004
Micrococcus sp., *Bacillus megaterium*	Gaskin and Bentham, 2005
Bacillus mycoides, *Corynebacterium* sp.	Gaskin and Bentham, 2005
Mycobacterium sp. strain HH1	Zhon *et al.*, 2006
Alcaligenes spp.	Gaskin and Bentham, 2005
Bacillus subtilis DM-04	Das and Mukherjee, 2007
Pseudomonas aeruginosa M	Das and Mukherjee, 2007
Bacillus cereus Py5	Lin and Cai, 2006
Bacillus megaterium Py6	Lin and Cai, 2006
Cyclolasticus sp. P1	Wang *et al.*, 2008
Pseudomonas spp. LP1, LP5, LP6	Obayori *et al.*, 2008
Mycobacterium spp. strains KA5, BS5 and KF4	Lease *et al.*, 2011
Proteus vulgaris 4Bi	Ceyhan, 2012

The organism was a Gram-negative and motile short rod. It produced yellowish circular colonies (2mm diameter) with odour like that of *E. coli* on Luria Bertani agar. It differed from the clinical strain in being galactose, maltose, cyclodextrin, histidine, phenyl amine, ampicillin and vancomycin positive, but rhamnose, cellubiose, acotinic acid, dextrin and inositol negative. Apart from pyrene, it also used catechol, naphthalene, fluorene and fluoranthene as sole carbon and energy source but not anthracene.

A pyrene-degrading isolate, *Saccharothrix xinjiangensis* strain PYX-6T, was obtained from a lake in China by Hu *et al.* (2004). The organism, an aerobic, Gram-positive filamentous bacterium, capable of growth on starch and a variety of carbohydrates, utilized anthracene, phenanthrene, pyrene and other aromatics as sole carbon and energy sources.

It degraded 14 percent pyrene supplied. Das and Mukherjee (2007) reported the degradation of pyrene by biosurfactant producing thermophilic strains of *Pseudomonas aeruginosa* and *Bacillus subtilis.*

Obayori *et al.* ((2008) reported three *Pseudomonas* strains utilizing pyrene as sole sources of carbon and energy. These organisms demonstrated versatility for a range of petroleum hydrocarbons and a few chlorinated aromatics. They were isolated from different hdrocabon polluted sites in Lagos, Nigeria.

Cometabolism and the Role of Consortium

Most of the research works on biodegradation of PAHs have been geared towards isolating, identifying and determining the degradative pathways of strains that use the substrates as sole carbon and energy sources. But it is a well established fact that in the natural environment, co-metabolism and interaction within complex communities play an immensely important role in the removal of contaminants from pristine environment (Bouchez *et al.*, 1995; Bastiaens *et al.*, 2000; Stach and Burns, 2002).

Thus cometabolism and the activities of consortia are crucial to the ultimate containment of persistent pollutants in the environment. Cometabolism can be defined as the fortuitous transformation of a compound in the presence of another metabolizable compound (Atlas, 1995). Enzymatically, it is a non-specific enzymatic reaction, with a substrate competing with a structurally similar primary substrate for the enzymes active site (Johnsen *et al.*, 2005). The effectiveness of co-metabolism in the removal of pollutants presupposes the interaction of two or more organisms. Such intricately interacting community capable of achieving degree of degradation which the individual populations can not achieve alone constitutes a consortium.

Mueller *et al.* (1989) showed that a seven-member culture capable of utilizing fluoranthene as a sole carbon and energy source was also able to cometabolize other high molecular weight PAHs including pyrene in a concentration range of 0.3-2.3 mg/l. One of the members of this consortium was later identified as a *Pseudomonas/Sphingomonas paucimobilis* EPA 505 (Mueller *et al.*, 1990). Weissenfels *et al.* (1990) documented the isolation of the soil microorganism, *Alcaligenes denitrificans* strain, which biodegraded fluoranthene at a rate of 0.3 μg/ml/day and cometabolized other PAHs including pyrene.

These observations raise question as to the implication of use of mixed culture in bioremediation efforts. Bouchez *et al.* (1995) employed six organisms, including two

Rhodococcus sp. strains capable of growth on pyrene and chrysene as sole sources of carbon to study the biodegradation of mixtures of PAHs by bacteria. It was found that all strains were capable of cometabolic inhibition and synergistic interactions. PAHs that were more water soluble than the PAH added originally were found to induce more inhibition. It was also observed that mineralisation yields were higher and biomass yields lower for high molecular weight PAHs than low molecular weight PAHs.

In an experiment on the cometabolism of pyrene by a *Sphingomonas* species strain KS14 growing on phenanthrene, Cho and Kim (2001) found that 62.5% of pyrene added was degraded by the bacterium within 8 days of incubation and the concentration of phenanthrene was coincidentally decreased to under the detection limit in this period. However, the degradation rate of phenanthrene as a growth substrate was lower than that of phenanthrene as sole carbon and energy source. Strain KS14 was able to degrade pyrene only when phenanthrene existed in the medium as a primary substrate. The authors, therefore, concluded that the enzyme produced by strain KS14 in order to degrade phenanthrene could degrade pyrene, and that pyrene alone could not induce such an enzyme, and furthermore that the lower degradation rate of primary substrate phenanthrene was the result of competition between primary substrate (phenanthrene) and secondary substrate (pyrene) for the same enzyme. Bacteria growing on pyrene have also been shown to cometabolize higher PAHs such as benzo(a)pyrene (Boonchan *et al.*, 1998). Kim *et al.* (2003b) showed that B*urkholderia* sp. strain 2A-12 was capable of fortuitous metabolism of pyrene when supplied with yeast extract or phenanthrene as co-substrate. The same was demonstrated for *Burkholderia* sp. strain VUN 10013 by Samtrakoon *et al.* (2008). *Stenotrophomonas maltophila* strain VUN 10013 was shown to degrade benzo(a)pyrene and benzo(a)anthracenene (10-15 mg/l) in 56 days when supplied with pyrene (250 mg/l) as co-substrate (Juhasz *et al.*, 1997).

Cometabolism has its own demerits. Important is the fact that there are so many organisms which although cannot use pyrene as sole carbon and energy sources but transform the PAH to metabolites which may interfere with the mineralisation of related PAHs in mixtures. Kazunga and Aitken (2000) examined the formation of non-mineral products of transformation of pyrene by four bacterial strains. *Pseudomonas stutzeri* strain P16 and *Bacillus cereus* strain P21 transformed pyrene primarily to *cis*-4,5-dihydroxypyrene (PYRdHD), the first intermediate in the pathway for aerobic bacterial mineralisation of pyrene. *Sphingomonas yanoikuyae* strain R1 transformed pyrene to PYRdHD and pyrene-4,5-dione (PYRQ). Both strain R1 and *Pseudomonas saccharophila* strain P15 transformed PYRdHD to PYRQ nearly stoichiometrically, suggesting that PYRQ is formed by oxidation of PYRdHD to 4,5-dihydroxypyrene and subsequent autooxidation to PYRQ. *Mycobacterium* strain PYR-1, also transformed PYRdHD to PYRQ at high initial concentrations of PHRdHD. However, strain PYR-1 was able to use both PYRdHD and PYRQ as growth substrate. PYRdHD was found to strongly inhibited phenanthrene degradation by strains P15 and R1 but had only a minor effect on strains P16 and P21. At their aqueous saturation concentrations both PYRdHD and PYRQ strongly inhibited benzo(a)pyrene mineralisation by strains P15 and R1. The authors concluded that products derived from pyrene transformation have the potential to accumulate in PAH-contaminated systems and that such products can affect the removal of other PAHs.

Molina *et al.* (1999) showed that *Mycobacterium* strain MR-1 mineralized pyrene and phenanthrene concurrently when pre-exposed to either of the two compounds, that is, showing evidence of cross-induction. Guha *et al.* (1999) examined the biodegradation of PAHs in mono-substrates and in mixture. In binary mixtures of naphthalene, phenanthrene and pyrene biodegradation kinetics were found to be comparable to sole-substrate experiments. In ternary mixtures, biodegradation of naphthalene was found to be inhibited. In another study, Sho *et al.* (2004) found that pyrene mineralization by *Mycobacterium* sp. strain S65 was enhanced by adding benzo(a)anthracene, benzo(a)pyrene or phenanthrene as co-substrates.

Pathways of Bacterial Metabolism of Pyrene

Metabolism of polycyclic aromatic hydrocarbons could either result in biotransformation/detoxification or in complete biodegradation or mineralization. The former often leads to change in structural complexity of chemical properties without utilization as sole source of carbon and energy. While in the latter case the degrader completely breaks down the substrates to CO_2 and H_2O deriving thereby carbon and energy. Since the commencement of the study of metabolism of pyrene by Heitkemp *et al.* (1988b), several of authors have identified metabolites denoting these two different trajectories (Walter *et al.*, 1991; Schneider *et al.*, 1996; Dean Ross and Cerniglia, 1996; Krivobok *et al.*, 2003; Kim *et al.*, 2005; Liang *et al.*, 2006; Kim *et al.*, 2007).

Whichever way, however, the initial step usually involves the addition of two atoms of oxygen to a benzene ring in the pyrene cluster leading to dihydroxylation which is the classic route of initiation of attack on the aromatic ring (Figure 2) and in a few cases cytochrome 450 mediated monohydroxylation heading to pyrenol (1-hydroxyl pyrene).

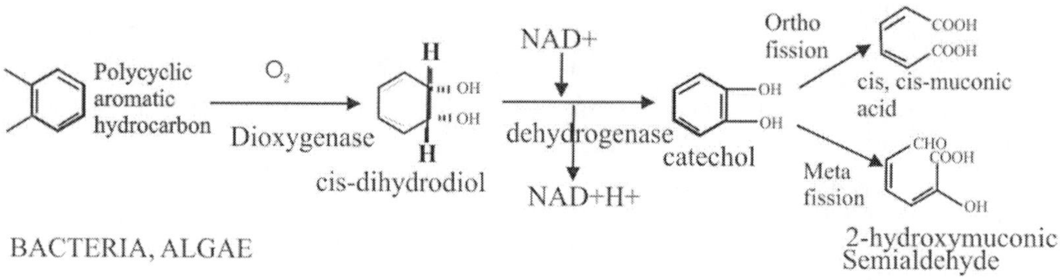

Figure 2. Classical pathway of dioxygenation and cleavage of aromatic ring by bacteria.

Till date, the classical routes of pyrene metabolism have been the ones described in *Mycobacterium* species, particularly *Mycobacterium vanbaalenii* PYR-1. Fundamentally, three routes are identifiable and these depend on the site of the initial attack:

1. Attack at the 4,5-carbon positions often lead to complete mineralization via O-phthalate. But depending on the culture condition, it also produces a number of dead end metabolites.
2. Another route that may be followed involves cleavage of the two central, i.e.K, rings. Usually after the cleavage of the ring at 4,5-position to produce 4,5-phenanthrene dicarboxylic acid, there is attack at the 9-10 position leading to hydroxylation and

subsequent formation of diphenic acid. This route leads to complete mineralization via O-phthalate but involves fewer metabolites.

3. When attack occurs at the 1,2- position there is usually biotransformation instead of complete mineralization, with methylated metabolites as the end products (Kim *et al.*, 2007; Kanaly and Harayama, 2010).

i. Classical pathways of complete mineralization of pyrene by *Mycobacterium* sp.

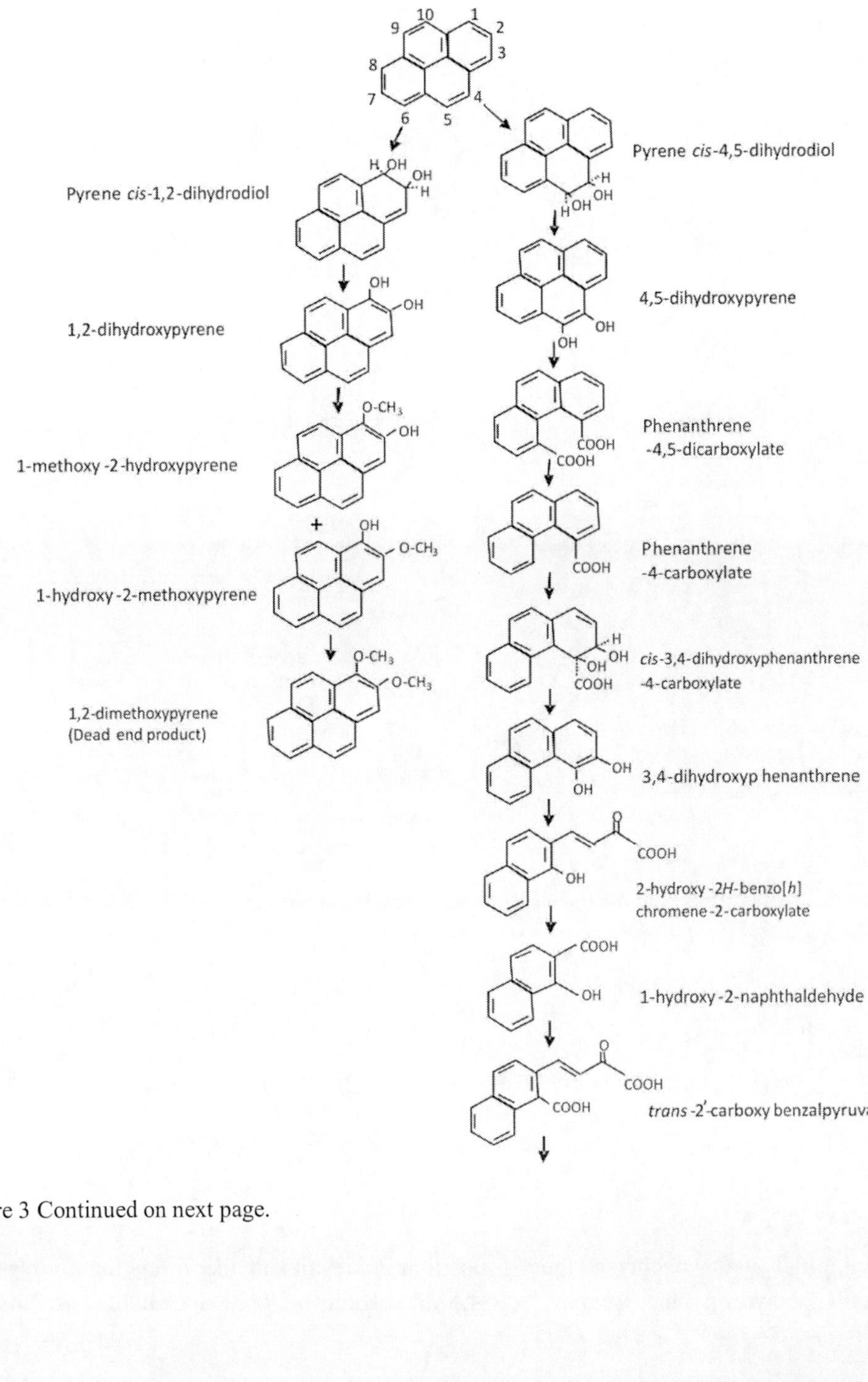

Figure 3 Continued on next page.

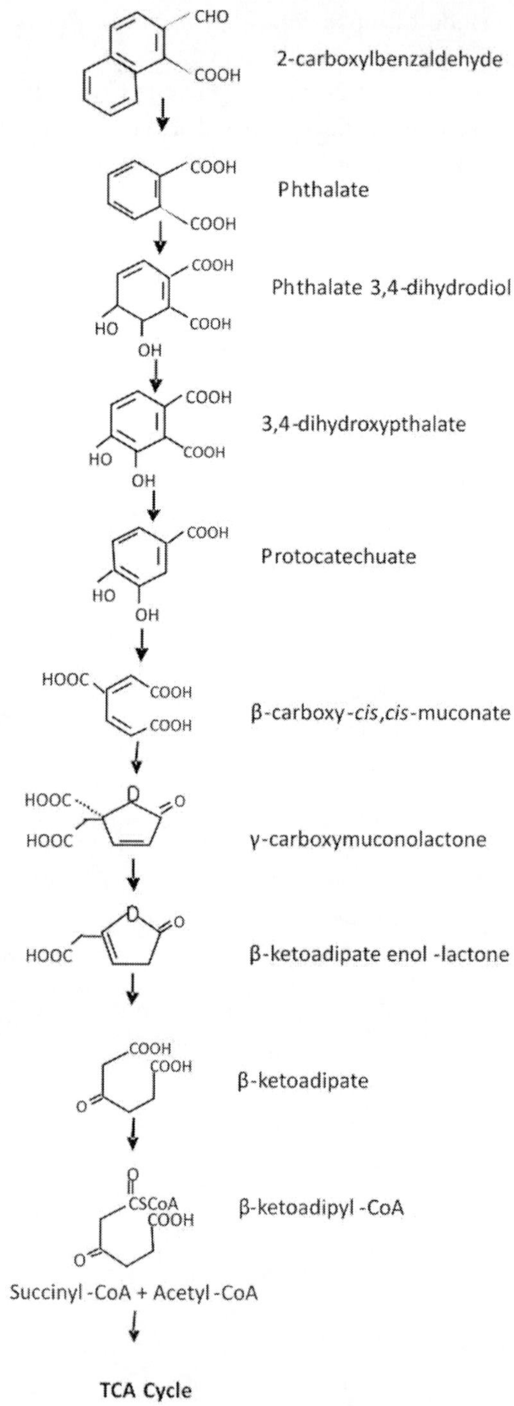

Figure 3. Classical pathways of pyrene degradation by *Mycobacterium* species (Krivobok *et al*., 2003; Kim *et al.,* 2007).

When initial attack occurs at the 4,5-position, it results in the formation of *cis*-4,5-dihydroxy-4,5-dihydropyrene (pyrene *cis*-4,5-dihydrodiol). This metabolite is further transformed by an NADH dependent dehydrogenase to form 4,5-dihydroxypyrene; the double hydroxylation compromises the integrity of the aromatic ring and prepares it for cleavage. A second dioxygenase enzyme, a cleavage dioxygenase mediates an intradiol (*ortho*) cleavage of the hydroxylated ring leading to the formation of phenanthrene 4,5-dicarboxylic acid (Heitkamp *et al.*, 1988b; Walter *et al.*, 1991; Dean-Ross and Cerniglia 1996; Schneider *et al.*, 1996; Rehman *et al.*, 1998; Krivobok *et al.*, 2003; Liang *et al.*, 2006) which is further decarboxylated to give phenanthrene 4-carboxylic acid (Heitkamp *et al.*, 1988b; Dean-Ross and Cerniglia, 1996; Schneider *et al.*, 1996). From this point onward, the route followed is similar to the pathway of phenanthrene degradation via O-phthalate (Krivobok *et al.*, 2003; Kim *et al.*, 2007). It is characterized by sequential oxidation and cleavage of the benzene rings up to O-phthalate and finally through β-keto adipate into the tricarboxylic acid cycle.

Phenanthrene-4-carboxylic acid is oxidized to *cis*-3,4-dihydroxy phenanthrene-4-carboxylic acid (which is unstable) and further oxidized to 3,4-dihydroxyphenanthrene. The dihydroxylated product undergoes *meta*- cleavage to yield 2-hydroxy-2,-4-benzo(b) chromeno-2-carboxylic acid; this product is hydrolyzed to 1-hydroxy-2-naphthaldohyde which is oxidized to 1-hydroxy-2-naphthenoic acid. This product undergoes *meta*- cleavage to yield *trans*-2-carboxybenzylpyruvate, which is further metabolized via 2-carboxylbenzaldehyde to Ophthalate. The above steps were fully described in *Mycobacterium* sp. strain 6PY1 by Krivobok *et al.* (2003) based on proteomic and genomic study are essentially the same as those described for *Mycobacterium vanbaaleni* PYR-1 by several authors including Kim *et al.* (2007).

The route from O-phthalate involves dehydroxylation at the 3,4- position to yield O-phthalate 3.4-dihydrodiol which is further metabolized via 3,4-dihydroxyphthalate to protocatechuate. Protocatechuate undergoes *ortho*-cleavage to yield β-carboxy *cis,cis* muconate, which is latonized to γ–carboxymucono lactone. This is isomerized to β-keto adipate enol-lactone, which finds its way into the TCA cycle via β-keto adipic acid, a key metabolite in the complete metabolism of aromatic compounds.

As shown in Figure 3 (Kim *et al.*, 2007), this pathway involves twenty seven steps. The first 12 steps occur between pyrene and O-phthalate, six from O-phthalate to protocatechuate and seven from protocatechuate to acetyl COA and succinyl CoA. Alternatively when trans dihydroxylation takes place leading to *trans*-4,5-dihydroxy-4,5-dihydrodiol, there is further transformation of this to *cis*-4,5-dihydroxy pyrene which again follows the route outlined above.

Quinone Oxidoreductase Mediated Side Reaction

4,5-dihydroxy pyrene can undergo non-enzymytic autooxidation to pyrene 4,5-dione. This quinone has been identified in a number of mycobacteria but has only been shown to be metabolized in *Mycobacterium* sp. strain KMS (Liang *et al.*, 2006). It was shown that the organism degraded pyrene-4,5-dione to phenanthrene 4,5-dicarboxylic acid and 4-phenanthroic acid which is further metabolized to phthalate as described above. It has been shown previously in *Mycobacterium* sp. strain PYR100 and *M. vanbaalenii* PYR-1 (Kim *et al.* 2003a; Kim *et al.*, 2004) that pyrene-4,5-dione can be reduced to 4,5 dihydroxy pyrene by a glucose reductase A similar enzyme was also identified in *Mycobacterium* sp. KMS the transformation of pyrene-4,5-dione by *Mycobacterium* sp. KMS is significant in that it suggests that the organism has mechanism to tolerate the toxicity of pyrene-4,5-dione and metabolize it to less toxic metabolites (*Kim et al*., 2006). This route may be considered as a side reaction as KMS uses essentially the same route as described for PYR-1 and *Mycobacterium* sp. 6PY1 above (Figure 3).

DEAD-END PRODUCTS FROM DIOXIDATION AT THE 4,5 POSITION

Meta-cleavage of pyrene-4,5-diol leads to the formation of 2-hydroxy-2-(phenanthrene-5-one-4-enyl) acetic acid, which is lactonized to 5-hydroxy-5H-4-oxa-yrene-5 carboxylic acid (Seo *et al.*, 2009).

Other dead-end products generated from metabolite of pathway following oxidation at 4,5s position include:

i. Phenanthrene 4-carboxylic acid methyl ester from phenanthrene 4-carboxylic acid;
ii. 4-hydroxy-phenanthrene from *cis*-3,4-hydroxyphenanthrene – 4-carboxylic acid and;
iii. 4-oxa phenanthrene-3-one from 3,4-dihydroxyphenenthrene

These four products were identified in PYR1 by Kim *et al.* (2005) and furthermore these authors further showed that PYR1 can metabolize phenanthrene-4,5-dicarboxylic acid from which a dead and product 4-oxa-pyrene-5-one is generated.

ALTERNATIVE ROUTE VIA OXIDATION OF THE TWO CENTRAL K-RINGS

A new pathway in the degradation of pyrene was opened when Vila *et al.* (2001) showed that in *Mycobacterium* sp. strain AP1 following formation of phenanthrene-4,5-dicarboxylic acid there is transformation to 5- hydroxyphenanthrene 4- carboxylic acid and dihyroxylation of the product of the 9,10-position loading to formation of 5,9, 10-dihydroxy-9, 10-dihydroxy phenanthrene-4-carboxylic acid.

Further oxidation leads to the production of a dihydroxylated product 5, 9, 10-trihydroxyphenanthrene -4- carboxylic acid. *Ortho*- cleavage as the 9, 10 position results in a metabolize 2-hydroxyl 2', 6, 6'-biphenyl dicarboxylic acid. This acid is metabolized to 6,6'-dihydroxy-2, 2' biphenyl dicarboxylic acid (diphenic acid) which is transformed to O-phthalate (Figure 4). In this pathway, which was more fully elucidated by Kanaly and Harayama (2010), one dead-end product is shown to be formed through the transformation of 5-hydroxyl phenanthrene-4-carboxylic acid. This dead-end product is 4-oxa-pyren-5- one (Kim *et al.*, 2005).

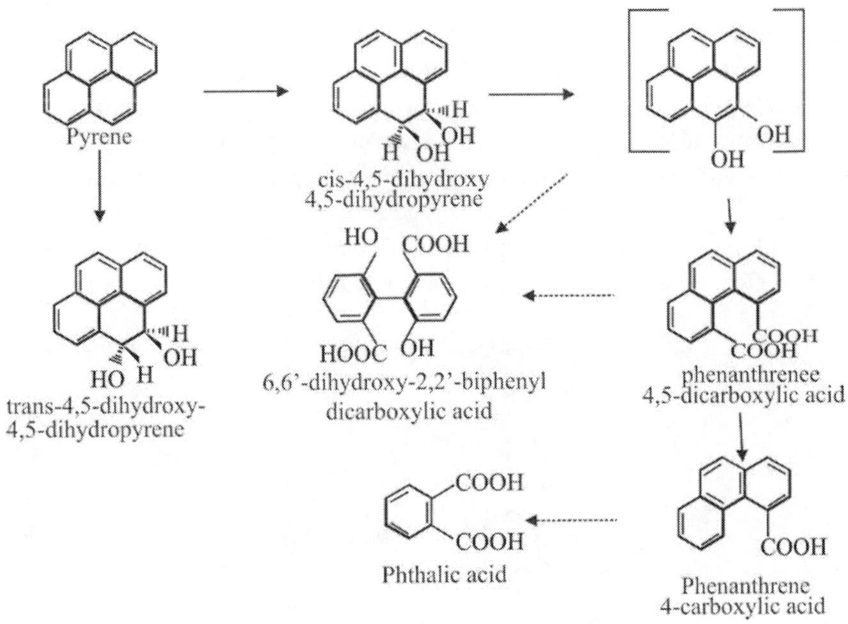

Figure 4. Schematic pathway proposed by Vila *et al.* (2001) for the degradation of pyrene by *Mycobacterium* sp. strain AP1.

TRANSFORMATION VIA ATTACK INITIATED AT THE 1,2-CARBON POSITION

Dioxygenation of pyrene at the 1, 2-C position leads to the formation of pyrene 1, 2-dihydrodiol. The dihydrodiol is further metabolized to pyrene 1, 2-diol via dehydrogenization. Further metabolism via *cis*-2-hydroxyl-3-(perinaphthenon-9yl)-propenic acid and 2-hydroxyl-2H-1-oxa- pyrene-2-carboxylic acid leads to 4- hydroxyperinaphthenone a dead-end product which was first identified as for back as 1988, by Heitkamp *et al.* (Seo *et al.*, 2009; Kanaly and Harayama, 2010).

Alternatively 1, 2-dihydroxyryrene is methylated at the 1-C and 2-C position to form a mixture of two products: 1-methoxy-2 hydroxyl pyrene and 1-hydroxyl-2-methoxy pyrene. These products are further methylated to 1, 2-dimethoxypyrene, a dead end product (Kim *et al.*, 2004; Kim *et al.*, 2005; Seo *et al.* 2009; Kanaly and Harayama, 2010). Thus, dioxygenation at the 1, 2-C position invariably leads to the formation of dead-end products and does not serve the organism as source of carbon or energy.

PATHWAYS OF PYRENE DEGRADATION BY BACTERIA OTHER THAN *MYCOBACTERIUM* SPECIES

Saccharothrix sp. PYX 6 has been shown to degrade pyrene via a pathway that involves catechol and benzyl propionic acid as they metabolize (Hu *et al.*, 2003). Based on this finding, the authors proposed the pathway shown in Figure 5. This shows that the organism shares a common upper pathway with *Mycobacterium* PYR-1 up to phenanthrenic acid before diverting. In the Gram-negative organism *Cycloclasticus* sp. strain P1 isolate from deep sea statement of the west pain tic degradation initiated by dihydroxylation at the 4, 5- position followed by transformation to cyclopenta (*def*) phenanthrenone (Wang *et al.*, 2008).

This is further metabolized through phenanthrenol, 1, 2-dihydroxy phenanthrenone and salicyclate to catechol which is then further processed via the *meta-* cleavage pathway to the tricarboxylic acid cycle (Figure 6). This was confirmed by the ability of cell free extract of the bacterium to cleave catechol to 2-hydroxymuconic semialdehyde.

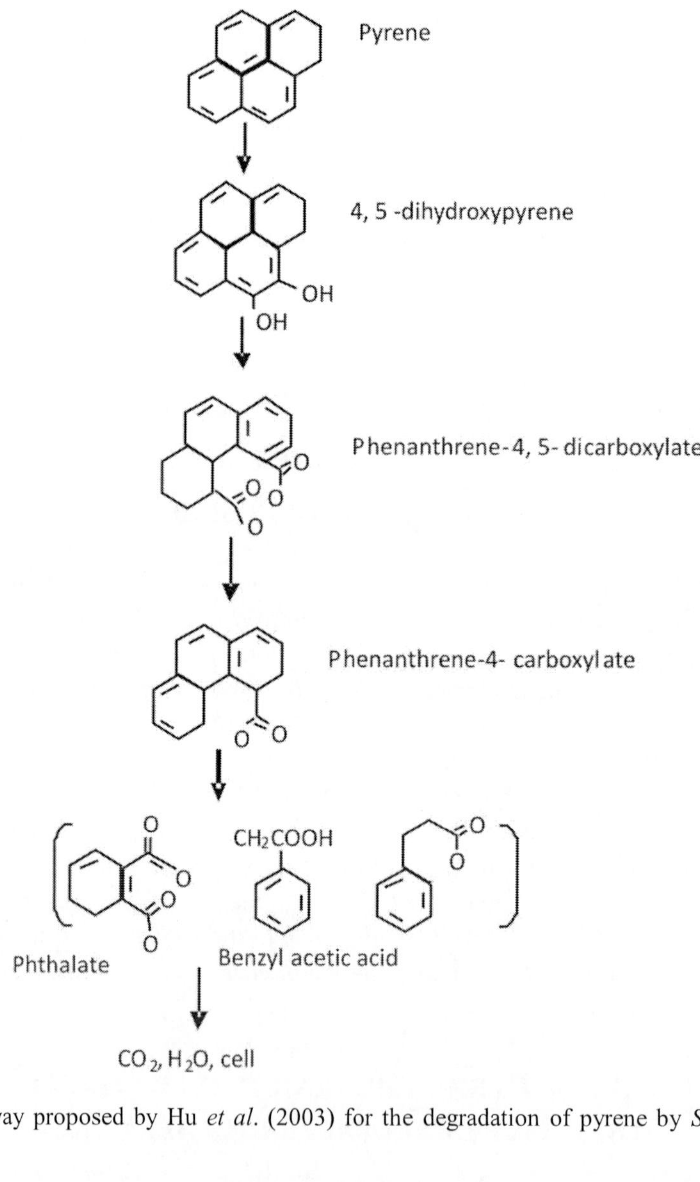

Figure 5. Pathway proposed by Hu *et al.* (2003) for the degradation of pyrene by *Saccharothrix* sp. PYX

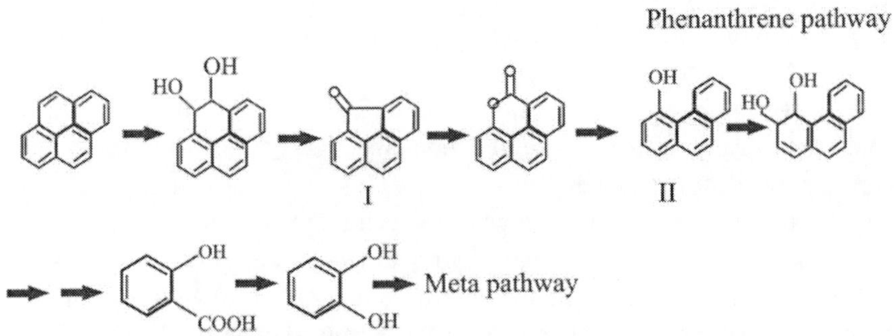

Figure 6. Pathway proposed for degradation of pyrene by *Cyloclasticus* sp. P1 (Wang *et al.,* 2008).

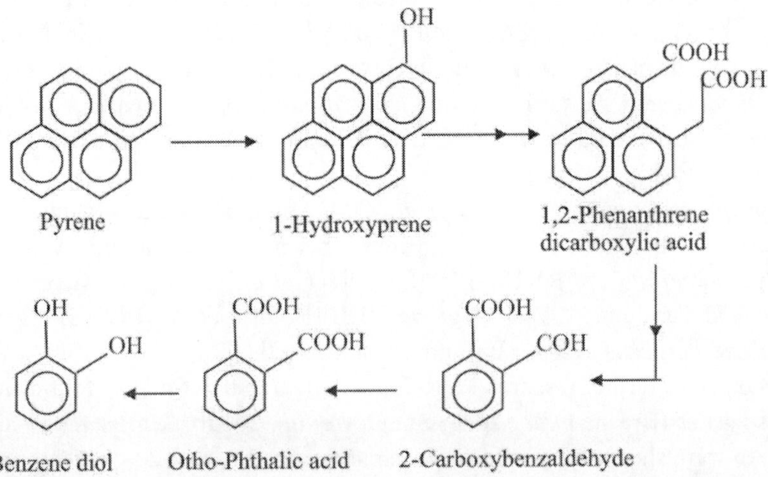

Figure 7. Pathway proposed for the degradation of pyrene by *L. adecarboxylata* PS4040 (Sarma *et al.,* 2010).

The identification of 1-hydroxypyrene as the ring hydroxylation product in metabolism of pyrene by Sarma *et al.* (2010) opened a new pathway in pyrene metabolism by bacteria as this particular metabolite had only been previously identified as product of oxidation by fungal enzymes or mammals. *Leclercia adecarboxylate* PS4040 was found to utilize a pathway that proceeded through 1-hydroxyl pyrene, 1, 2-phenanthrene dicarboxylic acid and 1, 2-benzene diol (catechol). All of these metabolites were isolated and conclusively identified by the authors. Figure 7 above shows the pathway proposed for pyrene degradation by *L. adecarboxylata.* In this pathway another unique metabolite is 1,2-phenanthrene dicarboxylic acid. The authors also went further to suggest that unlike in mycobacterial degradation of pyrene, this organism degraded pyrene without producing dead-end products.

GENES AND ENZYMES

Wang *et al.* (2000) cloned the first gene thought to be involved in pyrene metabolism by *Mycobacterium vanbaaleni* PYR-1. The gene, identified as *Kat*G, encodes the enzyme catalase-peroxidase, an 81-KDa a protein, suggested to be involved in protection of dioxygenases from oxidative inactivation by exogenous oxidants or removal of endogenous hydrogen peroxide that may have been generated as intermediate during PAH metabolism. The gene clustered with catalase peroxidase proteins of other *Mycobacterium* species. Although not specifically a pyrene degradation specific gene, this opened the door to investigation of pyrene specific genes.

Dioxygenation or addition of two atoms of oxygen to an aromatic ring is usually the first step in degradation of polycyclic aromatic hydrocarbons. Thus process is catalyzed by an enzyme belonging to a family of aromatic ring hydroxylating dioxygenase. They are multi-component enzyme systems consisting of an electron transport chain and a terminal dioxygenase composed of large (α) and small (β) subunits (Mason and Cammack, 1992).

The dioxygenase genes involved in pyrene degradation were first cloned by Khan *et al.* (2001). The gene *nid*A was found to be distantly related to well known ring hydroxylating dioxygenase genes such as *nah*Ac, *ndo*Ba and *phn*AC genes found in *Pseudomonas* and *Bukholderia* spp. It was, however found to cluster highly with dioxygenases such as *nid*A and *ph*A from *Rhodococcus* sp. strain 124 and *Nocardioides* sp. strain Kp7. Also cloned were the dioxygenase small β) subunit (*nid*B) and the aldehyde dehydrogenase (*nid*D) genes and a partial sequence of the *nid*C gene.

Subsequent investigation (Brezna *et al.*, 2003) by other authors showed that the genes have conserved homologous sequences in four other mycobacteria known to degrade pyrene namely *Mycobacterium* sp. RJG11-135 (Schneider *et al.*, 1996), *M. flavescens* PYR-GCK (Dean-Ross and Cerniglia, 1996), *M. gilvum* BBI (Boldrin *et al.*, 1993; Bottger *et al.*, 1997) and *M. frederiksbergense* FAn9^T (Willumsen *et al.*, 2001).

In the course of further research PYR-1 was shown to have at least four different terminal dioxygenase genes (Brezna *et al.*, 2003; Stingley *et al.*, 2004b; Kim *et al.*, 2006). The *nid*AB and *pht*AaAb were shown to transform phenanthrene and pyrene to phenanthrene and pyrene *cis* dihydrodiol (Khan *et al.*, 2001; Stingley *et al.*, 2004b) and phthalate into phthalate *cis* 3,4 dihydrodiol (Stingley *et al.*, 2004a). The fourth copy dioxygenase gene *nid*A3B3 was described by Kim *et al.*, (2006). The protein designated as nidA3AB can oxidize several polycyclic aromatic hydrocarbons (PAHs) including fluoranthene, pyrene and benzo (a) anthracene to *cis*-dihydrodiols with the aid of ferredoxin (*phd*C) and ferredoxin reductase (*phd*D) genes. Two dioxygenases involved in pyrene degradation where identified by Krivobok *et al.* (2003) from *Mycobacterium* sp. 6PY1. These two enzymes were designated Pdo1 and pd02. Pdo1 was found to catalyze the dihydroxylation of pyrene and phenanthrene while pdo2 was found to catalyze dihydroxylation of only phenanthrene. The pdo1 protein components were found to have 98.5% match with the nidA protein (α-subunit arene dioxygenase) from PYR-1 and 99% match with the nidB protein (β-subunit of arene dioxygenase from the same organism. Furthermore the authors isolated in all 22 pyrene-induced proteins, half of which were similar to phenanthrene induced proteins found in the PAH degrading *Nocardioides* sp. strain KP7. Enzymes with the following functions were identified: thiosulfate sulfur transferase, trans- carboxylbenzalpyruvate hydratase aldolase, hydratase aldolase, 1 hydroxy-2-naphthoate dioxygenase and aldehyde dehydrogenase. The second β-subunit of arene dioxygenase had 62% match with phdB protein found in strain KP7, the hydratase-aldolase 72% with phdG, while the aldehyde dehydrogenase had 90% similarity with phdH protein found in KP7. The genes encoding the protein pd01 were cloned and identified as pdoB1 (β-subunit) and pdoA1 (α-subunit). These genes were markedly different from the *nid*BA gene from strain PYR-1 in that they are not inducible by naphthalene. The genes encoding the second dioxygenase were also cloned and designated *pdo*A2 and *pdo*B2. The dioxygenase genes involved in pyrene degradation by *Mycobacterium* sp. strain 6PY1 were localized in two loci. Based on this proteomic study, the authors proposed a metabolic pathway containing six steps specific to pyrene degradation, two specific to phenanthrene and seven common to both pyrene and phenanthrene (i.e. form 3,4 dihydroxyphenanthrene to O-phthalic acid). Sho *et al.* (2004) cloned the dioxygenase genes involved in pyrene degradation by *Mycobacterium* sp. strain S65 and found that the genes for pyrene degradation have unique organization in this organism and are located in two different loci. Two large (α) subunits pdoA/x and mdA/x were discerned.

Table 3. Enzymes involved in the metabolism of pyrene by *Mycobacterium vanbaalenii* PYR-1 and the steps they catalyze

Metabolic function	Matching protein	Enzymatic step
Pyrene/phenanthrene ring-hydroxylating	NidA PdoA1	Pyrene → pyrene *cis* 4,5-dihydrodiol
Pyrene/phenanthrene ring-hydroxylating oxygenase, β subunit	PdoA1 NidB2	Pyrene → pyrene *cis* 4,5-dihydrodiol
Dihydrodiol dehydrogenase	PhdE	Pyrene *cis*-4,5-dihydrodiol →4,5-dihydroxypyrene, *cis* 3,4-dihydroxyphenanthrene 4-carbocxylate →3,4dihydroxyphenanthrene Pyrene *cis* 1,2-dihydrrodiol→ 1,2 dihydroxy pyrene
Ring cleavage dioxygenase	PhdF	4,5-dihydroxypyrene→phenanthrene 4,5 dihydroxy carboxylate 3,4 dihydroxyphenanthrene→ 2-hydroxy-2,4-benzo(h) chromene-2-carboxylate
Decarboxylase	PhtC	Phenanthrene -4-carboxylate→ *cis* 3,4-dihydroxyphenanthrene -4-carboxylate 3,4-dihydroxyphthalate→ protocatechuate
Phenanthrene ring-hydrolaxylating oxygenase, α-subunit	PdoA2	Phenanthrrene-4-carboxylate → *cis* 3,4-dihydroxyphenanthrene-4-carboxylate
Phenanthrene ring-hydrolaxylating oxygenase, β-subunit	PdoB2	Phenanthrrene-4-carboxylate → *cis* 3,4-dihydroxyphenanthrene-4-carboxylate
Hydratase-aldolase	PhdG PhdG	2-hyhroxy-2,4-benzo(h)chromene-2-carboxylate → 1-hydroxy-2-naphthaldehyde
Aldehyde dehydrogenase	NidD PhdH	1-hydroxy-2-naphthaldehyde → 1-hydroxy-2-naphthoate
1-hydroxy-2-naphthoate dioxygenase	PhdI PhdI	1-hydroxy-2-naphthoate→ *trans*-2-carboxybenzalpyruvate
Trans-2-carboxybenzalpyruvate hydratase-aldolase	PhdJ PhdJ	*Trans*-2-carboxybenzalpyruvate → 2-carboxylbenzaldehyde
2-carboxylbenzaldehyde dehydrogenase	PhdK	2-carboxylbenzaldehyde → phthalate
Phthalate 3,4-dioxygenase, α-subunit	PhtAa PhtA1	Phthalate → phthalate 3,4-dihydrodiol
Phthalate 3,4-dioxygenase, β-subunit	PhtAb PhtA2	Phthalate → phthalate 3,4-dihydrodiol

Table 3. (Continued)

Metabolic function	Matching protein	Enzymatic step
Oxygenase ferredoxin component	PhtAc PhtAc	Pyrene → pyrene *cis* 4,5-dihydrodiol 1 Phenanthrrene-4-carboxylate → *cis* 3,4-dihydroxyphenanthrene-4-carboxylate Phthalate → phthalate 3,4-dihydrodiol Pyrene → pyrene *cis* 1,2, dihydrodiol
Oxygenase reductase component	PhtAd PhtAd	Pyrene → pyrene *cis* 4,5-dihydrodiol 1 Phenanthrrene-4-carboxylate → cis 3,4-dihydroxyphenanthrene-4-carboxylate Phthalate → phthalate 3,4-dihydrodiol Pyrene → pyrene *cis* 1,2, dihydrodiol
Phthalate 3,4-dihidrodiol dehydrogenase	PhtB PhtB	Phthalate 3,4-dihydrodiol → 3,4 dihydroxy phthalate
Metabolic function	Matching protein	Enzymatic step
Protocatechuate 3,4-dioxygenase, α-subunit	PcaG	Protocatechuate → β-carboxy-*cis,cis*-muconate
Protocatechuate 3,4-dioxygenase, β-subunit	PcaH	Protocatechuate → β-carboxy-*cis,cis*-muconate
β-carboxy-*cis, cis*-muconate cycloisomerase	PcaB	β-carboxy-*cis,cis* muconate → γ-carboxymuconolactone
γ-carboxymuconolactone decarboxylase/β-ketoadipate	PcaL	γ-carboxymuconolactone → β-ketoadipate β-ketoadipate enol lactone → β-ketoadipate
β-ketoadipate succinyl-CoA transferase, α-subunit	PcaI	β-ketoadipate → β-ketoadipyl/ CoA
β-ketoadipate succinyl-CoA transferase, β-subunit	PcaJ	β-ketoadipate → β-ketoadipyl/ CoA
β-ketoadipyl-CoA thiolase	PcaF	β-ketoadipyl CoA → succinyl CoA and Acetyl CoA
Fluoranthene/pyrene ring hydroxylating oxygenase, α-subunit	NidA3	Pyrene → pyrene *cis* 4,5-dihydrodiol Pyrene → pyrene *cis* 1,2, dihydrodiol
Fluoranthene/pyrene ring hydroxylating oxygenase, β-subunit	NidB3	Pyrene → pyrene *cis* 4,5-dihydrodiol Pyrene → pyrene *cis* 1,2, dihydrodiol
Catechol O-methyltransferase	MT1743	1,2-dihydroxypyrene → 1 methoxy-2-hydroxy-nydrxypyrene 1-methoxyl-2-hydroxypyrene

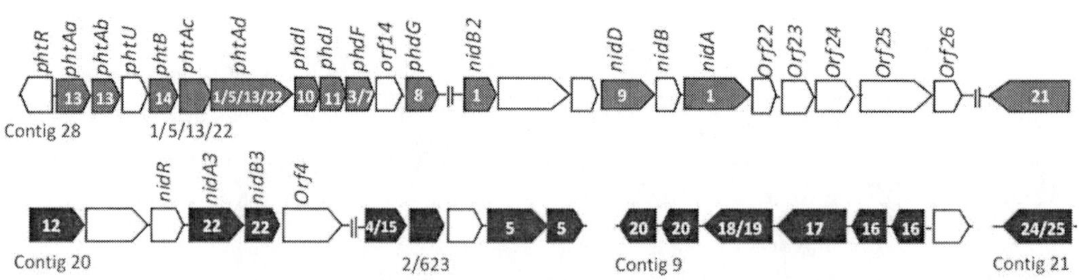

Figure 8. Organization of the gene clusters involved in the catabolism of pyrene in *M. vanbaalenii* PYR-1. Genes/ORFs are represented by arrows. The gene/ORFs labeled in italics are those also identified in other organisms. The gene products identified as proteins involved in pyrene degradation are represented by shaded arrows with numbers that correspond to enzymatic steps in the pathway shown in Fig. 3 (Adapted from Kim *et al.*, 2007).

The *nid*A was 99.5% similar to the nidA of *Mycobacterium vanbaaleni* strain PYR-1 and the *pdo*A was 89% similar to PYR-1 *nid*A. A small subunit (β) (*pdo*A) was found to have 89% identity with the *nid*B gene found in PYR 1. Furthermore, the two loci were also found to contain two alcohol dehydrogenase genes *pdo*C/H and *nid*C/H just downstream of *nid*A and *pdo*H and one unknown open reading fame (orf) – orfp6 and orfN4).

In the locus where the *pdo* was found an IS3-like tansposase was also found upstream of *pdo*B. Building on results of previous works in the data base, Kim *et al.* (2007) used genomic sequence analysis to identify genes involved in pyrene degradation and equally used proteome analyses of wells exposed to pyrene to identify proteins involved in pyrene degradation. In all, 27 enzymes involved in complete degradation of pyrene to the TCA cycle were identified. The enzymes include 14 responsible for the degradation of pyrene to phthalate, responsible for the degradation of phthalate to protocatechuate of responsible for the pathway from protocatechuate to acetyl-COA and succinate-COA and 3 responsible for the transformation of pyrene to 1,2-dimethoxypyrene. These enzymes functions and matching proteins together with the steps they catalyse are shown in Table 3 above.

Corresponding genes to the 27 enzymes involve in pyrene degradation have all been identified in *Mycobacterium vanbbaalenii* PYR-1 (Kim *et al.*, 2007). The 27 genes/ORTs were found to be organized into four contigs (Figure 8). Usually genes involved in the degradation of PAHs have been found to be clustered together uniquely in the genome (Saito *et al.*, ;2000; Habe *et al.*, 2003; Pinyakong *et al.*, 2003; Habe *et al.*, 2005). The organization of the genes for pyrene degradation is shown to follow this same rule. It was shown that 12 of the 27 genes involved in pyrene degradation were also involved in phenanthrene degradation (Stingley *et al.*, 2004b) while Krivobok *et al.* (2003) identified the genes involved in pyrene dioxygenation.

INVOLVEMENT OF PLASMIDS

Hydrocarbon degradation has been shown to be mediated by enzymes encoded by genes encoded on both plasmids and chromosomes. Whereas reports abound on the involvement of plasmids in the degradation of low molecular weight PAHs, until recently there were only very few reports on plasmid mediated catabolism of HMW PAHs (Obayori and Salam, 2010). Genes involved in pyrene catabolism by such versatile organisms as *Mycobacterium vanbaalenii* PYR-1 and *Mycobacterium* sp. 6PY1 have been found to be localized on chromosome (Heitkamp *et al.*, 1988; Khan *et al.*, 2001; Krivobok *et al.*, 2003; Kim *et al.*, 2007). Recently, however, interesting findings have been reported. Sho *et al.* (2004) isolated a *Mycobacterium* sp. strain S65 from a set fuel contaminated site. This isolate, which was able to grow on pyrene, phenanthrene and fluoranthene as sole carbon and energy sources, was found to have a large plasmid. But the *nidA* gene found on the plasmid was not associated with pyrene degradation.

In a landmark report, Miller *et al.* (2007) demonstrated plasmids in *Mycobacterium* sp. KMS (150 - 225 kb) *Mycobacterium* sp. MC (150-215 kb) and *Mycobacterium* sp. PYR-GCK (300 - 350 kb) *Mycobacterium* sp. JLS and the well studied *Mycobacterium vanbaaleni* PYR-1 possessed no plasmid. The authors speculated that there are consensus sequences in the nidBA promoters that are involved in the regulation of expression of these genes. The dioxygenase involved in pyrene degradation was highly conserved. Lin and Cai (2008) reported the isolation of plasmid from a pyrene degrading consortium made up of two strains, *Bacillus cereus* Py5 and *Bacillus megaterium* Py6. Pyrene degradation was localized on the plasmid after plasmid transformed *E. coli* was shown to degrade pyrene.

However, there are no reports yet on the involvement of plasmids in the degradation of pyrene by pseudomonads. Obayori *et al.* (2008) could not also detect plasmids in the three pyrene degrading *Pseudomonas* strains isolated from polluted tropical sites in Lagos, Nigeria in spite of the remarkable versatility of these isolates.

Given that some isolates which had been previously taken for granted as non-plasmid bearers are now known to posses plasmids, it could be conjectured that the choice of method is very important in detection of plasmids, as most large plasmids are likely to escape detection using routine techniques used for small plasmids. It is not unlikely that plasmids are more involved in the degradation of HMW PAH than previously thought.

ENVIRONMENTAL FACTORS AFFECTING BIODEGRADATION OF PYRENE

According to Cerniglia (1992), microbial degradation of PAHs in aquatic and terrestrial environments is influenced strongly by a number of factors which include temperature, pH, soil type, salinity, aeration, nutrients among others.

TEMPERATURE

The rate of petroleum hydrocarbon degradation has generally been observed to decrease, with decreasing temperature; this has been attributed primarily to decreased rate of enzymatic activity (Shiaris, 1989; Leahy and Colwell, 1990). Most of the mycobacteria that have been found to degrade pyrene grow best at ambient or room temperature (24 - 30°C) and are inhibited by high temperature of >42°C. However, organisms capable of degrading PAHs under low temperature have been reported (Eriksson *et al.*, 2003). But a few thermophilic pyrene degraders have been reported (Fietkenhauer *et al.*, 2003; Das and Mukherjee, 2005; Zeinali *et al.*, 2007). Thus, it would appear that effect of temperature on pyrene degradation is directly a function of intrinsic bacterial physiology than PAH degradation property.

EFFECT OF PH

Soil pH is highly variable and could range from 2.5 to 11.0. Most bacteria and fungi are however favoured by values near neutrality. Thus, small pH shifts have been found to have profound effects of the degradation of pyrene both in pure culture and in the environment. Grosser *et al.* (1991) observed that a pyrene-degrading *Mycobacterium* sp. mineralised pyrene more rapidly in acidic (pH 4.4) soil without significant lag time, compared with slightly alkaline soil (pH 7.8). It has been shown that the mycolic acid layer, the primary permeability barriers against hydrophobic molecules, may be disrupted under acidic conditions (Kim *et al.*,

2005). The pH of the media and biotransformation buffer was found to affect the growth rate of *M. vanbaaleni* PYR-1 and its degradation of phenanthrene and pyrene. The substrate could get into the cell at a higher rate at pH 6.5 compared with pH 7.5. Furthermore, the organism was found to produce higher levels of O-methylated derivatives of non-K-region pyrene diol at pH 6.5. Pyrene-degrading mycobacteria appear to be favoured by acidic conditions while pseudomonads are favoured by slightly alkaline soils (Uyttebroek *et al.*, 2007). Acidity generally decreases microbial degradation activity (Koivula *et al.*, 2004). However, mineralisation of pyrene in soil has been reported at pH of 3.6. Roper and Pfaender (2001) showed the mineralisation of pyrene in mineral soil of hardwood forest at pH 4.8.

Also, pyrene degradation was found to be optimal at pH of 9 for *Mycobacterium* sp. strain MPH-1 an alkaliphilic bacterium isolated from an arid environment in Japan.

Effect of Aeration

Biodegradation of most organic pollutants occurs at a faster rate under aerobic conditions (Crawford, 2002) and anaerobic conditions have even been found to lower PAH removal in sediments than lowering temperature (Eriksson *et al.*, 2003). This is because the initial step in the catabolism of aromatic hydrocarbons involves the oxidation of the substrate by oxygenases, for which molecular oxygen is required (Leahy and Colwell, 1990; Cerniglia, 1992). In an experiment to demonstrate the effect of oxygen tension on the rate of biodegradation of PAHs, Hurst *et al.* (1998) found the degradation rates of ^{14}C-pyrene and PAH compounds to be enhanced under soil gas oxygen concentrations between 2% and 21% in the contaminated soil Between 45% and 55% of ^{14}C-pyrene spiked onto the soil was mineralized after 70 days at soil gas oxygen levels between 2% and 21%. No statistically significant mineralization was found to occur at 0% oxygen concentrations.

Salinity

Salinity plays a very important role in determining the biodegradation of PAHs including pyrene by specialized strains in soil (Jimenez and Bartha, 1996). Shiaris (1989) observed a positive correlation between the rates of mineralization of naphthalene and phenanthrene and salinity in estuarine sediments. But in culture, extreme salinity levels in excess of 5% have been found to be limiting (Heitkamp *et al.*, 1988a). Kastner *et al.* (1998) demonstrated that the salinity of the inoculating medium could be a very important factor in survival of bacteria applied as inoculum during bioremediation. Obayori *et al.* (2008) reported three pyrene degrading *Pseudomonas* strain that tolerated NaCl concentrations of more than 3%.

Presence of Other Organic Matters

The presence of organic substrates or PAHs also affects the rate of degradation of pyrene. Other organic substances present in water or soil may serve as more readily available sources of nutrient for soil microorganisms to the exclusion of attack on more refractory substances such as pyrene and other hydrocarbons. Egli (2002) noted that the formation of enzymes catalysing the degradation of organic pollutants may be repressed when the cells grow on other substrates, but the constitution background level of expression is often sufficient for immediate consumption of the pollutant if it becomes available in low amounts (Johnsen *et al.* 2005). Conversely, such substances may serve as good sources of cofactors that could ginger the organisms, giving the initial kick needed to metabolise recalcitrant substances. They may also exact other not very specific influences. Humic acid is one of such substances that have been well-studied (Larsson *et al.*, 1988; Seibel *et al.,* 1996; Haderlein *et al.*, 2001; Fava *et al.*, 2004). It has been demonstrated that humic acid enhanced the aqueous solubility of PAHs in petroleum products, but further studies have shown that high concentration of humic acid in excess of 50 mg/l inhibited pyrene degradation. Klinge *et al.* (2001) reported that sludge amendment enhanced biodegradation of pyrene in soil with concomitant increase in total heterotrophic bacteria, protozoa and pyrene mineralizing bacteria populations. Furthermore, increased pyrene degradation has been reported in the presence of corn steep liquor (Obayori *et al.*, 2010).

INORGANIC NUTRIENTS

Microorganisms generally require mineral (inorganic) nutrients sources for growth. If any of the required nutrients is lacking or becomes limiting, particularly the macromineral elements, pollutant degradation reaction may be slow even though carbon and energy sources required for growth are available (Giordani *et al.*, 1998; Lehtola *et al.*, 1998; Vidali, 2001). Hydrocarbons generally are weak sources of nutrients such as nitrogen and phosphorus which are essential building blocks of structural macromolecules like proteins and nucleic acids as well as enzymes and coenzymes. They are therefore highly limiting in petroleum polluted environments (Leahy and Colwell, 1990). Thus there is usually a sudden depletion of these nutrients in oil polluted environments (Andrew and Jackson, 1996).

Jones *et al.* (2008) found that the addition of nitrogen to either continuously mixed slurry or field-wet soil incubations increased the rate and extent of mineralisation of pyrene. In static microcosm it was also discovered that nitrogen amendment decreased the lag period before significant pyrene mineralization occurred.

The demand for micro- and macronutrients including growth factor depend on the species or strain. Whereas the pyrene degrader *Mycobacterium* PYR-1 (Heitkamp *et al.*, 1988a) had requirement for growth factor, *Mycobacterium* CH1 (Churchill *et al.*, 1999) had no such growth factor requirements. Vitamin solution, yeast extract, peptone and yeast extracts are presently some of the growth supplements in use (Heitkamp *et al.*, 1988a; Kastner *et al.*, 1994; Jimenez and Bartha, 1996).

SURFACTANTS

There are considerable reports in the literature on the effect of synthetic surfactants on the biodegradation of pyrene (Tiehm, 1994; Tiehm and Fritzsche, 1995; Thibault *et al.*, 1996). Jimenez and Bartha (1996) reported increase in pyrene mineralization of up to 154 %

following addition of detergent below critical micelle concentration (cmc). Paraffin oil, squalone, tridecetylcyclohexane and cis-9'-tricosene (0.8 %, v/v) were found to double pyrene mineralization. However, the more hydrophobic synthetic surfactants that seem to exhibit better solubilizing properties are tolerated by only a few bacteria.

Recent report (Das and Mukherjee, 2007) showed that biosurfactants secreted by three strains, namely: *Bacillus subtilis* DM-OX and *Pseudomonas aeruginosa* M and NM, enhanced the apparent solubility of pyrene and influenced the bacterial cell surface hydrophobicity resulting in higher uptake of pyrene by bacteria. Also, biosurfactant production was found to enhance the degradation of pyrene by *Pseudomonas fluorescens* strain 29L (Husain, 2008).

CONCLUSION

Pyrene is by nature a highly recalcitrant polycyclic aromatic hydrocarbon whose resistance to biodegradation is enhanced by its peri-condensed structure. However, in recent years bacteria capable of mineralizing pyrene in pure culture have been reported. Generally, these organisms degrade pyrene by sequential cleavage of the rings after dihydroxylation. The lower pathways of complete mineralization are similar to the pathways of degradation of lower PAHs like phenanthrene. The genes encoding the enzymes have been found to be borne on chromosome or plasmids, but largely on the former. Furthermore, many of these enzymes have broad specificity for other polyaromatic hydrocarbons. The role of biosurfactants is yet to be fully established, but fact abound that manipulation of environmental factors and nutrient addition could increase the rate of bacterial metabolism of pyrene. This holds a high prospect for bioremediation of polluted environmental compartments.

REFERENCES

Andrew, R.W.J. and Jackson, J.M. (1996). Pollution and waste management, In: *The Natural Environment and Human Impact*. Longman Publishers, Singapore. Pp. 281-297.

Atlas, R.M. (1995). *Principles of Microbiology*. Mosby, St. Louis. P. 808.

Balaban, A. T., Schleyer P. V. R. and Rzepa, H. S. (2005). Crocker, not Armit and Robinson, begat the six aromatic electrons. *Chem. Rev.*105: 3436-3447.

Bastiaens, L., Springael, D., Wattiau, P., Harms, H., deWachter, R., Verachtert, H., and Diels, L. (2000). Isolation of adherent polycyclic aromatic hydrocarbon (PAH)-degrading bacteria using PAH-sorbing carriers. *Appl. Environ. Microbiol.* 66: 1834–1843.

Boldrin, B., Tiehm, A., and Fritzsche, C. (1993). Degradation of phenanthrene, fluorene, fluoranthene, and pyrene by a *Mycobacterium* sp. *Appl. Environ. Microbiol.* 59: 1927–1930.

Boonchan, S., Britz, M.L., and Stanley, G.A. (1998). Surfactant-enhanced biodegradation of high molecular weight polycyclic aromatic hydrocarbons by *Stenotrophomonas maltophila*. *Biotechnol. Bioeng.* 59: 482–494.

Böttger, E.C., Kirschner, P., Springer, B., and Zumft, W. (1997). Mycobacteria degrading polycyclic aromatic hydrocarbons. *Int. J. Syst. Bacteriol.* 47: 247.

Bouchez, M., Blanchet, D., Bardin, V., Haeseler, F., and Vandecasteele, J.-P. (1999). Efficiency of defined strains and of soil consortia in the biodegradation of polycyclic aromatic hydrocarbon (PAH) mixtures. *Biodegradation* 10: 429–435.

Bouchez, M., Blanchet, D. and Vandecasteele, J.-P. (1995). Degradation of polycyclic aromatic hydrocarbons by pure strains and by defined strain associations: inhibition phenomena and cometabolism. *Appl. Microbiol. Biotechnol.* 43: 156-164.

Brezna, B., Khan, A.A., and Cerniglia, C.E. (2003). Molecular characterization of dioxygenases from polycyclic aromatic hydrocarbon-degrading *Mycobacterium* spp. *FEMS Microbiol. Lett.* 223: 177–183.

Cerniglia, C. (1992). Biodegradation of polycyclic aromatic hydrocarbons. *Biodegradation* 3: 351–368.

Cerniglia, C.E. and Shuttleworth, K.L. (2002) Methods for Isolation of Polycyclic Aromatic Hydrocarbons (PAHs) – Degrading Microorganism and Procedures for Determination of Biodegradation Intermediates and Environmental monitoring of PAHs. In: *Manual of Environmental Microbiology*, 2nd edition (C.J Hurst, R.L Crawford, G.R Knudson, M.J MClnerny L.D Stetzenbach, eds). ASM Press, Washington D.C pp. 972 – 986.

Ceyhan, N. (2012). Biodegradation of pyrene by a newly isolated *Proteus vulgaris*. *Sci. Res. Essays*. 7:66-77.

Cheung, P.Y., and Kinkle, B.K. (2001). *Mycobacterium* diversity and pyrene mineralization in petroleum-contaminated soils. *Appl. Environ. Microbiol.* 67: 2222–2229.

Churchill, S.A., Harper, J.P., and Churchill, P.F. (1999). Isolation and characterization of a *Mycobacterium* species capable of degrading three- and four-ring aromatic and aliphatic hydrocarbons. *Appl. Environ. Microbiol.* 65: 549–552.

Crawford, R.L. (2002). Overview: biotransformation and biodegradation. In: *Manual of Environmental Microbiology*, 2nd Edition, (C.J. Hurst, R.C. Crawford, G.R. Kausdsen, M.J. McInerny and L.D. Stetzenbach , eds.) ASM Press, Washington, D.C. Pp.972-986.

Das, K., and Mukherjee, A.K. (2007). Differential utilization of pyrene as the sole source of carbon by *Bacillus subtilis* and *Pseudomonas aeruginosa* strains: role of biosurfactants in enhancing bioavailability. *J. Appl. Microbiol.* 102: 195–203.

Das, K., and Mukherjee, A.K. (2005). Characterization of biochemical properties and biological activities of biosurfactants produced by *Pseudomonas aeruginosa* mucoid and non-mucoid strains isolated from hydrocarbon contaminated soil samples. *Appl. Microbiol. Biotechnol.* 69:192-199.

Dean-Ross, D., and Cerniglia, C.E. (1996). Degradation of pyrene by *Mycobacterium flavescens*. *Appl. Microbiol. Biotechnol.* 46: 307–312.

Derz, K., Klinner, U., Schuphan, I., Stackebrandt, E., and Kroppenstedt, R.M. (2004). *Mycobacterium pyrenivorans* sp. nov., a novel polycyclic-aromatic-hydrocarbondegrading species. *Int. J. Syst. Evol. Microbiol.* 54: 2313–2317.

Egli, T. (2002). Microbial degradation of pollutants at low concentrations and in the presence of alternative carbon substrates: emerging patterns In: *Focus on Biotechnology. Biotechnology for environment: Strategy and Fundamentals*, Vol 3A. (S.N. Agathos, and W. Reineke, eds.), Khiwer Academic Publishers, Dordrecht, pp. 131-139.

Eriksson, M., Sodersten, E., Yu, Z., Dalhammar, G., and Mohn, W.W. (2003). Degradation of polycyclic aromatic hydrocarbons at low temperature under aerobic and nitrate-reducing conditions in enrichment cultures from Northern soils. *Environ. Microbiol.* 69: 275–284.

Fava, F., Berselli, S., Conte, P., Piccolo, A., and Marchetti, L. (2004). Effects of humic substances and soya lecithin on the aerobic bioremediation of a soil historically contaminated by polycyclic aromatic hydrocarbons (PAHs). *Biotechnol Bioeng* 88: 214–223.

Feitkenhauer, H., Müller, R., and Märkl, H. (2003). Degradation of polycyclic aromatic hydrocarbons and long chain alkanes at 60–70°C by *Thermus* and *Bacillus* spp. *Biodegradation* 14: 367–372.

Garcia-Cruz, I and Martinez-Magadan, J.M. (2007). DFT molecular dynamics study of pyrene radical species. *Petrol. Sc. Technol*. 25: 67-80.

Gaskin, S., and Bentham, R. (2005). Comparison of enrichment methods of the isolation of pyrene-degrading bacteria. *Int. Biodeterior. Biodeg.* 56: 80–85.

Giordani, G., Donnelly, A. and Azzoni, R. (1998). The uptake of inorganic phosphate by Z. noltii in the Basin d' Arcachon, In: *Proceedings (Handbook) of Summer Conference of the Society for Applied Microbiology*, University of Lancaster, UK. Pp. 22-23.

Grosser, R. J., Warshawsky, D. and Vestal, J.R. (1991). Indigenous and enhanced mineralization of pyrene, benzo[*a*]pyrene, and carbazole in soils. *Appl. Environ. Microbiol.* 57:3462–3469.

Guha, S., Peters, C.A. and Jaffe, P.R. (1999). Multisubstrate biodegradation kinetics of naphthalene, phenanthrene and pyrene mixtures. *Biotechnol. Bioeng*. 65: 491-499.

Habe, H., Chung, J.S., Ishida, A., Kasuga, K., Ide, K. Takemura, T. Nojiri, H. Yamane, H. and Omori, T. (2005). The fluorene catabolic linear plasmid in*Terrabacter* sp. strain DBF63 carries the β-ketoadipate pathway genes, *pcaRHGBDCFIJ*, also found in proteobacteria. *Microbiology* 151:3713–3722.

Habe, H., Kanemitsu, M., Nomura, M., Takemura, T., Iwata, K., Nojiri, H., Yamane, H. and Omori, T. (2004). Isolation and characterization of an alkaliphilic bacterium utilizing pyrene as a carbon source. *J. Biosci. Bioeng.* 98: 306–308.

Habe, H., Miyakoshi, M. , Chung, J. , Kasuga, K., Yoshida,Y., Nojiri, H. and Omori. T. (2003). Phthalate catabolic gene cluster is linked to the angular dioxygenase gene in *Terrabacter* sp. strain DBF63. *Appl. Microbiol. Biotechnol.* 61:44–54.

Haderlein, A., Legros, R., and Ramsay, B. (2001). Enhancing pyrene mineralization in contaminated soil by the addition of humic acids or composted contaminated soil. *Appl. Microbiol. Biotechnol.* 56: 555–559.

Heitkamp, M.A., Franklin, W., and Cerniglia, C.E. (1988a). Microbial metabolism of polycyclic aromatic hydrocarbons: isolation and characterization of a pyrene-degrading bacterium. *Appl. Environ. Microbiol.* 54: 2549–2555.

Heitkamp, M.A., Freeman, J.P., Miller, D.W., and Cerniglia, C.E. (1988b). Pyrene degradation by a *Mycobacterium* sp.: identification of ring oxidation and ring fission products. *Appl. Environ. Microbiol.* 54: 2556–2565.

Hilyard, E.J., Jones-Meehan, J.M., Spargo, B.J. and Hill, R.T. (2008). Enrichment, isolation, and phylogenetic identification of polycyclic aromatic hydrocarbon-degrading bacteria from Elizabeth River Sediments *Appl. Environ. Microbiol.*74:1176–1182.

Hu, Y.-T., Zhou, P.-J., Zhou, Y.-G., Liu, Z.-H., and Liu, S.-J. (2004). *Saccharothrix xinjiangensis* sp. nov., a pyrene degrading actinomycete isolated from Tianchi Lake, Xinjiang, China. *Int. J. Syst. Evol. Microbiol.* 54: 2091–2094.

Hu, Y.-T., Ren, F., Zhou, P., Xia, M. and Liu, S.. (2003). Geradation of pyrene and characterization of *Saccharothrix* sp. PYX-6 from the oligotrophic Tianchi Lake in Xinjiang Uygur Autonomous Region, China. *Chinese Sci. Bull.* 48(20): 2210-2215.

Hurst, C.J., Sims, R. C., Sims, J.L., Sorensen, D.L., Mclean, J.E. and Huling, S. (1998). polycyclic aromatic hydrocarbon biodegradation as a function of oxygen tension in contaminated soil. In: Proceedings of the 10th Annual Conference on Hazardous waste Research pp 20 – 32.

Husain, S. (2008). Effect of surfactants on pyrene degradation by *Pseudomonas fluorescens* 29L. *World J. Microbiol. Biotechnol.* 24: 2411–2419.

Jimenez, I. Y. and Bartha, R. (1996). Solvent-augmented mineralisation of pyrene by a *Mycobacterium* sp. *Appl. Environ. Microbiol*. 62:2311-2316.

Johnsen, A.R., Wick, L.Y. and Harms, H. (2005). Principles of microbial PAH-degradation in soil. *Environ. Poll.* 133:71-84.

Jones, M.D., Singleton, D.R., Carstensen, D.P., Powell, S.N., Swanson, J.S., Pfaender, F.K., and Aitken, M.D. (2008). Effect of incubation conditions on the enrichement of pyrene-degrading bacteria identified by stable-isotope probing in an aged, PAH-contaminated soil. *Microb. Ecol.* 56: 341–349.

Juhasz, A.L., Britz, M.L., and Stanley, G.A. (1997). Degradation of fluoranthene, pyrene, benz[*a*]anthracene and dibenz[*a,h*]anthracene by *Burkholderia cepacia*. *J. Appl. Microbiol.* 83: 189–198.

Kanaly, R. A. and Harayama, S. (2010). Advances in the field of high-molecular-weight polycyclic aromatic hydrocarbon biodegradation by bacteria. *Microbial Biotechol.* 3(2): 132-164.

Kanaly, R.A., and Harayama, S. (2000). Biodegradation of high-molecular-weight polycyclic aromatic hydrocarbons by bacteria. *J. Bacteriol.* 182: 2059–2067.

Kästner, M., Breuer-Jammali, M., and Mahro, B. (1994). Enumeration and characterization of the soil microflora from hydrocarbon-contaminated soil sites able to mineralize polycyclic aromatic hydrocarbons (PAH). *Appl. Microbiol. Biotechnol.* 41: 267–273.

Kästner, M., Breuer-Jammali, M., and Mahro, B. (1998). Impact of inoculation protocols, salinity, and pH on the degradation of polycyclic aromatic hydrocarbons (PAHs) and survival of PAH-degrading bacteria introduced into soil. *Appl. Environ. Microbiol.* 64: 359–362.

Kazunga, C., and Aitken, M.D. (2000) Products from the incomplete metabolism of pyrene by polycyclic aromatic hydrocarbon-degrading bacteria. *Appl. Environ. Microbiol.* 66: 1917–1922.

Kelley, I., Freeman, J.P., Evans, F.E., and Cerniglia, C.E. (1993). Identification of metabolites from the degradation of fluoranthene by *Mycobacterium* sp. strain PYR-1. *Appl. Environ. Microbiol.* 59: 800–806.

Khan, A.A., Wang, R.-F., Cao, W.-W., Doerge, D.R., Wennerstrom, D., and Cerniglia, C.E. (2001). Molecular cloning, nucleotide sequence, and expression of genes encoding a polycyclic aromatic ring dioxygenase from *Mycobacterium* sp. strain PYR-1. *Appl. Environ. Microbiol.* 67: 3577–3585.

Khan, A.A., Kim, S.-J., Paine, D.D., and Cerniglia, C.E. (2002). Classification of a polycyclic aromatic hydrocarbonmetabolizing bacterium, *Mycobacterium* sp. strain PYR-1, as *Mycobacterium vanbaalenii* sp. nov. *Int. J. Syst. Evol. Microbiol.* 52: 1997–2002.

Kim, Y.-H., Engesser, K.-H., and Cerniglia, C.E. (2003). Two polycyclic aromatic hydrocarbon o-quinone reductases from a pyrene-degrading *Mycobacterium*. *Arch. Biochem. Biophys.* 416: 209–217.

Kim, Y.-H., Freeman, J.P., Moody, J.D., Engesser, K.H., and Cerniglia, C.E. (2005). Effects of pH on the degradation of phenanthrene and pyrene by *Mycobacterium vanbaalenii* PYR-1. *Appl. Microbiol. Biotechnol.* 67: 275–285.

Kim, S.-J., Jones, R.C., Cha, C.-J., Kweon, O., Edmondson, R.D., and Cerniglia, C.E. (2004a). Identification of proteins induced by polycyclic aromatic hydrocarbon in *Mycobacterium vanbaalenii* PYR-1 using two-dimensional polyacrylamide gel electrophoresis and *de novo* sequencing methods. *Proteomics* 4: 3899–3909.

Kim, S.-J., Kweon, O., Freeman, J.P., Jones, R.C., Adjei, M.D., Jhoo, J.-W., Edmondson R.D. and Cerniglia, C. E. (2006). Molecular cloning and expression of genes encoding a novel dioxygenase involved in low- and high-molecular weight polycyclic aromatic hydrocarbon degradation in *Mycobacterium vanbaalenii* PYR-1. *Appl. Environ. Microbiol.* 72: 1045–1054.

Kim, S.-J., Kweon, O., Jones, R.C., Freeman, J.P., Edmondson, R.D., and Cerniglia, C.E. (2007). Complete and integrated pyrene degradation pathway in *Mycobacterium vanbaalenii* PYR-1 based on systems biology. *J. Bacteriol.* 189: 464–472.

Kim, T.J., Lee, E.Y., Kim, Y.J., Cho, K.S., and Ryu, H.W. (2003b). Degradation of polyaromatic hydrocarbons by *Burkholderia cepacia* 2A-12. *World J. Microbiol. Biotechnol.* 19: 411–417.

Kim, Y.-H., Moody, J.D., Freeman, J.P., Brenzna, B., Engesser, K.-H., and Cerniglia, C.E. (2004b). Evidence for the existence of PAH-quinone reductase and catechol-*O*methyltransferase in *Mycobacterium vanbaalenii* PYR-1. *J. Ind. Microbiol. Biotechnol.* 31: 507–516.

Klinge, C., Gejlsbjerg, B., Ekelund, F. and Madsen, T. (2001). Effects of sludge-amendment on mineralization of pyrene and microorganisms in sludge and soil. *Chemosphere* 45:625–634.

Krivobok, S., Kuony, S., Meyer, C., Louwagie, M., Willison, J.C., and Jouanneau, Y. (2003). Identification of pyreneinduced proteins in *Mycobacterium* sp. strain 6PY1: evidence for two ring-hydroxylation dioxygenases. *J. Bacteriol.* 185: 3828–3841.

Koivula, T.T., Salkinoja-Salonen, M., Peltola, R. and Romantschuk M. (2004). Pyrene degradation in forest humus microcosms with or without pine and its mycorrhizal fungus. *Environ. Qual.* 33: 46-53.

Larsson, P., Okla, L., and Tranvik, L. (1988). Microbial degradation of xenobiotic, aromatic pollutants in humic water. *Appl. Environ. Microbiol.* 25:1864-1867.

Leahy, J. G., and Colwell, R.R. (1990). Microbial degradation of hydrocarbons in the environment. *Microbiol. Rev.* 54:305–315.

Lease, C.W.M., Bentham, R.H., Gaskin, S.E., and Juhasz, A.I. (2011). Isolation and identification of pyrene mineralizing *Mycobacterium* spp. From contaminated and uncontaminated soils. *Appl. Environ. Soil Science.* Doi:10.1155/2011/409643.

Lehtola, M., Miettinen, I., Vartianen, T. and Martikainen, P. J. (1998). Purification of drinking water phosphorus and bacterial growth. In: Proceedings (Handbook) of Summer Conference of the Society for Applied Microbiology, University of Lancaster, UK. pp. 16-17.

Liang, Y., Gardner, D.R., Miller, C.D., Chen, D., Anderson, A.J., Weimer, B.C., and Sims, R.C. (2006). Study of biochemical pathways and enzymes involved in pyrene degradation by *Mycobacterium* sp. strain KMS. *Appl. Environ. Microbiol.* 72: 7821–7828.

Lin, Y., and Cai, L.-X. (2008). PAH-degrading microbial consortium and its pyrene-degrading plasmids from mangrove sediment samples in Huian, China. *Mar. Pollut. Bull.* 57: 703–706.

Mason, J. R., and Cammack. R. (1992). The electron-transport proteins of hydroxylating bacterial dioxygenases. *Annu. Rev. Microbiol.* 46:277–305.

Miller, C.D., Hall, K., Liang, Y.N., Nieman, K., Sorensen, D., Issa, B., Anderson, A. J., Sims, R.C. (2004). Isolation and characterization of polycyclic aromatic hydrocarbon-degrading *Mycobacterium* isolates from soil. *Microbial Ecol.* 48: 230–238.

Miller, C.D., Child, R., Hughes, J.E., Benscai, M., Der, J.P., Sims, R.C., and Anderson, A.J. (2007). Diversity of soil *Mycobacterium* isolates from three sites that degrade polycyclic aromatic hydrocarbons. *J. Appl. Microbiol.* 102: 1612–1624.

Molina, M., Araujo, R., and Hodson, R.E. (1999). Crossinduction of pyrene and phenanthrene in a *Mycobacterium* sp. isolated from polycyclic aromatic hydrocarbon contaminated river sediments. *Can. J. Microbiol.* 45: 520–529.

Moody, J.D., Freeman, J.P., Doerge, D.R., and Cerniglia, C.E. (2001). Degradation of phenanthrene and anthracene by cell suspensions of *Mycobacterium* sp. strain PYR-1. *Appl. Environ. Microbiol.* 67: 1476–1483.

Moody, J.D., Fu, P.P., Freeman, J.P., and Cerniglia, C.E. (2003). Regio- and stereoselective metabolism of 7,12- dimethylbenz[*a*]anthracene by *Mycobacterium vanbaalenii* PYR-1. *Appl. Environ. Microbiol.* 69: 3924–3931.

Moody, J.D., Freeman, J.P., Fu, P.P., and Cerniglia, C.E. (2004). Degradation of benzo[*a*]pyrene by *Mycobacterium vanbaalenii* PYR-1. *Appl. Environ. Microbiol.* 70: 340–345.

Moody, J.D., Freeman, J.P., and Cerniglia, C.E. (2005). Degradation of benz[*a*]anthracene by *Mycobacterium vanbaalenii* strain PYR-1. *Biodegradation* 16: 513–526.

Mueller, J. G., Chapman, P. J., Blattmann, B. O. and Pritchard, P.H. (1990). Isolation and characterization of a fluoranthene-utilizing strain of *Pseudomonas paucimobilis*. *Appl. Environ. Microbiol.* 56: 1079-1086.

Mueller, J. G., Chapman, P. J., and Pritchard, P.H. (1989). Action of a fluoranthene-utilizing bacteria from geographically diverse soils. *Antonie Van Leeuwenhoek* 71: 392-343

Obayori, O.S. and Salam, L.B. (2010). Degradation of polycyclic aromatic hydrocarbons: Role of plasmids. *Sci. Res. and Essays*. 5(25) 4093-4106.

Obayori, O.S., Ilori, M.O., Adebusoye, S.A., Oyetibo, G.O., and Amund, O.O. (2008). Pyrene-degradation potentials of *Pseudomonas* species isolated from polluted tropical soils. *World J. Microbiol. Biotechnol.* 24: 2639–2646.

Obayori, O. S., Ilori, M. O., Adebusoye, S. A., Oyetibo, G. O., Omotayo, A. E. and Amund O. O. (2010). Effects of corn steep liquor on growth rate and pyrene degradation *by Pseudomonas* strains. *Curr. Microbiol.* 60:407-411.

Pinyakong, O., Habe, H. and Omori, T. (2003). The unique catabolic genes in sphingomonads degarading polycyclic aromatic hydrocarbons (PAHs) *J. Gen. Appl. Microbiol.* 49:1-19.

Rehmann, K., Noll, H.P., Steinberg, C.E.W., and Kettrup, A.A. (1998). Pyrene degradation by *Mycobacterium* sp. Strain KR2. *Chemosphere* 36: 2977–2992.

Roper, J.C., and. Pfaender, F.K. (2001). Pyrene and chrysene fate in surface soil and sand microcosms. *Environ. Toxicol. Chem.* 20:223–230.

Saito, A., Iwabuchi, T., and Harayama, S. (2000). A novel phenanthrene dioxygenase from *Nocardiodes* sp. Strain KP7: expression in *Escherichia coli*. *J. Bacteriol.* 182: 2134–2141.

Sarma, P.M., Bhattacharya, D., Krishnan, S., and Lal, B. (2004). Degradation of polycyclic aromatic hydrocarbons by a newly discovered enteric bacterium, *Leclercia adecarboxylata*. *Appl. Environ. Microbiol.* 70: 3163–3166.

Schneider, J., Grosser, R., Jayasimhulu, K., Xue, W., and Warshawsky, D. (1996). Degradation of pyrene, benz[*a*]anthracene, and benzo[*a*]pyrene by *Mycobacterium* sp. strain RJGII-135, isolated from a former coal gasification site. *Appl. Environ. Microbiol.* 62: 13–19.

Seo, J-S, Keum, Y-S, and Li, Q.X. (2009). Bacterial degradation of aromatic compounds. *Int. J. Environ. Res. Public Health.* 6:278-309.

Shiaris, M.P. (1989). Seasonal biotransformation of naphthalene, phenanthrene and benzo(a)pyrene in surficial estuarine sediments. *Appl. Environ. Microbiol* 55:1391-1399.

Sho, M., Hamel, C., and Greer, C.W. (2004). Two distinct gene clusters encode pyrene degradation in *Mycobacterium* sp. strain S65. *FEMS Microbiol. Ecol.* 48: 209–220.

Somtrakoon, K., Suanjit, S., Pokethitiyook, P., Kruatrachue, M., Lee, H., and Upatham, S. (2008). Phenanthrene stimulates the degradation of pyrene and fluoranthene by *Burkholderia* sp. VUN10013. *World J. Microbiol. Biotechnol.* 24: 523–531.

Stach, J.E.M., and Burns, R.G. (2002). Enrichment versus biofilm culture: a functional and phylogenetic comparisonof polycyclic aromatic hydrocarbon-degrading microbial communities. *Environ. Microbiol.* 4: 169–182.

Stingley, R.L., Brezna, B., Khan, A.A., and Cerniglia, C.E. (2004a). Novel organization of genes in a phthalate degradation operon of *Mycobacterium vanbaalenii* PYR-1. *Microbiology* 150: 3749–3761.

Stingley, R.L., Khan, A.A., and Cerngilia, C.E. (2004b). Molecular characterization of a phenanthrene degradation operon of *Mycobacterium vanbaalenii* PYR-1. *Biochem. Biophys. Res Commun.* 322: 133–146.

Thibault, S.I., Anderson, M. and Frankenberger Jr., W.I. (1996). Influence of surfactants on pyrene desorption and degradation in soil. *Appl. Environ. Microbiol.* 62: 283-287.

Tiehm, A and Fritzsche, C. (1995). Utilization of solubilized and crystalline mixtures of polycyclic aromatic hydrocarbons by a *Mycobacterium* sp. *Appl. Microbiol. Biotechnol.* 42: 964 – 968.

Tiehm, A. (1994). Degradation of polycyclic aromactic hydrocarbons in the presence of synthetic surfactants. *Appl. Environ. Microbiol.* 60: 258 -263.

USDHS (1990). Toxicological Profile for Polycyclic Aromatic Hydrocarbons. US Dept. of Health and Human Services, Washington D.C.

Uyttebroek, M., Veimeir, S., Wattiu P., Ryngaert A., and Springael, D. (2007). Characterisation of cultured Enriched from acidic polycyclic aromatic hydrocarbon contaminated soil for growth on pyrene at low pH. *Appl. Environ. Microbiol.* 73(10): 3159 – 3164.

Vidali, M. (2001). Bioremediation. An overview. *Pure Appl. Chem.* 73: 1163-1172.

Vila, J., López, Z., Sabaté, J., Minguillón, C., Solanas, A.M., and Grifoll, M. (2001). Identification of a novel metabolite in the degradation of pyrene by *Mycobacterium* sp.

Strain AP1: actions of the isolate on two- and three-ring polycyclic aromatic hydrocarbons. *Appl. Environ. Microbiol.* 67: 5497–5505.

Walter, U., Beyer, M., Klein, J., and Rehm, H.-J. (1991). Degradation of pyrene by *Rhodococcus* sp. UW1. *Appl Microbiol Biotechnol* 34: 671–676.

Wang, B., Lai, Q., Cui, Z., Tan, T., and Shao, Z. (2008). A pyrene-degrading consortium from deep-sea sediment of the West Pacific and its key member *Cycloclasticus* sp. *Environ Microbiol* 10: 1948–1963.

Wang, R.-F., Wennerstrom, D., Cao, W.-W., Khan, A.A., and Cerniglia, C.E. (2000). Cloning, expression, and characterization of the *katG* gene, encoding catalase-peroxidase, from the polycyclic aromatic hydrocarbon-degrading bacterium *Mycobacterium* sp. strain PYR-1. *Appl. Environ. Microbiol.* 66: 4300–4304.

Weissenfels, W. D., Beyer, M. and Klein, J. (1990). Degradation of phenanthrene, fluorene and fluoranthene by pure bacterial cultures. *Appl. Microbiol. Biotechnol.* 32:479–484.

Willumsen, P., Karlson, U., Stackebrandt, E., and Kroppenstedt, R.M. (2001). *Mycobacterium frederiksbergense* sp. nov., a novel polycyclic aromatic hydrocarbon-degrading *Mycobacterium* species. *Int. J. Syst. Evol. Microbiol.* 51: 1715–1722.

Zhang, H., Kallimanis, A., Koukkou, A.I., and Drainas, C. (2004). Isolation and characterization of novel bacteria degrading polycyclic aromatic hydrocarbons form polluted Greek soils. *Appl. Microbiol. Biotechnol.* 65: 124–131.

Zeinali, M., Vossoughi, M., and Ardestani, S.K. (2007). Characterization of a moderate thermophilic *Nocardia* species able to grow on polycyclic aromatic hydrocarbons. *Lett. Appl. Microbiol.* 45: 622–628.

Zhong, Y., Luan, T., Zhou, H., Lan, C., and Tam, N.F.Y. (2006). Metabolite production in degradation of pyrene alone or in a mixture with another polycyclic aromatic hydrocarbon by *Mycobacterium* sp. *Environ. Toxicol. Chem.* 25: 2853– 2859.

In: Pyrene
Editors: Petr Ruzicka and Tomas Kral

ISBN: 978-1-62417-291-5

Chapter 4

LOW-TEMPERATURE LASER-INDUCED FLUORESCENCE STUDIES OF VARIOUS BIOLOGICAL SYSTEMS WITH PYRENE CHROMOPHORE

***Chen Lin*[1], *Ximao Feng*[1] *and Ryszard Jankowiak*[1,2,*]**
[1]Department of Chemistry and
[2]Department of Physics, Kansas State University,
Manhattan, KS, US

ABSTRACT

Pyrene (Py) and various metabolites with a Py chromophore are powerful probes often used in biophysical and biochemical sciences due to their high fluorescent quantum yields and relatively long fluorescence lifetimes. Their fluorescent properties are very sensitive to the change in their immediate surroundings and are used to probe their local environment. In this chapter, we review recent applications using low- and high-resolution fluorescence-based methodologies (i.e., the non-line-narrowing (NLN) and fluorescence line-narrowing (FLN) spectroscopies) that utilize their electronic properties. These methodologies are used in conformational studies of biomolecules containing Py chromophores and in identification of enantiomeric metabolites containing Py moiety. We demonstrate this approach allows probing the ground- and excited-state vibrational frequencies of various metabolites and DNA adducts, and can provide insight into the binding site to DNA bases in various *in vitro* and *in vivo* experiments. The position of the fluorescent origin-band and the so called "fingerprint" of molecule(s) based on their characteristic vibrational frequencies can also provide insight on the mechanism of DNA adduct formation. It is also shown that laser-based fluorescence spectroscopies can successfully probe interactions of metabolites with monoclonal antibodies, including cross-reactivity and conformational multiplicity.

[*] Corresponding author; email: ryszard@ksu.edu.

1. INTRODUCTION

Pyrene (Py) belongs to the category of polycyclic aromatic hydrocarbons (PAH) consisting of four fused benzene rings. This chromophore was first isolated from coal tar [1] and could be found as a product of incomplete combustion under a wide range of conditions [2]. Benzo[*a*]pyrene (B[*a*]P) has five condensed aromatic rings. This molecule is the most studied carcinogenic PAH present in a cigarette smoke, and is known to induce skin, lung, and stomach tumors in laboratory animals [3]. One metabolic pathway of B[*a*]P is the activation by cytochrome P450 (P450) 1A1/1B1 and epoxide hydrolase to its final electrophilic derivatives, e.g., 7,8-dihydroxy-9,10-epoxy-7,8,9,10-tetrahydro-B[*a*]P (BPDE), followed by the epoxide moiety of BPDE electrophilically attacking the exocylic amino group of guanine to produce diastereomeric N^2-guanine lesions (i.e., N^2-BPDE-dG adduct). Note that BPDE-derived DNA adducts also have Py chromophores. These lesions are bulky, so they can block major replicative DNA polymerases, and lead to G→T transversions and G→A transition mutations [4-7]. Though BPDE itself is nearly not fluorescent (i.e., it has a very low fluorescence quantum yield), BPDE-derived DNA adducts (upon binding to guanine and/or adenine [8, 9]), as well as their hydrolysis products (i.e., benzo[*a*]pyrene tetraols (BPTs)), have pyrene-like fluorescence due to the planar structure of the highly fluorescent pyrenyl moiety, although their fluorescent origin band is red-shifted in comparison with Py. Furthermore, fluorescence spectra of these molecules are known to be extremely sensitive to the molecular structure and polarity of its microenvironment [10, 11].

Py is known to have a long fluorescence lifetime (τ_M^{fl} = 450 ns) and relatively high quantum yield (e.g., $\Phi_M = 0.60$ in cyclohexane); subscripts M stand for monomer. However, conventional fluorescence measurement can only provide information about (0,0)-transition shifts and intensity variation of the origin and vibronic bands affected by solvent. On the contrary, low-temperature, high-resolution laser-induced fluorescence spectroscopy is more informative due to the vibrationally resolved spectra. In this approach, so-called fluorescence line-narrowing spectroscopy (FLNS) [10-19], contributions from thermal and solvent broadening could be largely reduced or even eliminated, resulting in highly structured fluorescence spectra with information about ground-state and excited-state vibrational energy levels. In addition, this "information-rich" detection technique can be combined with various separation methods, e.g., high-performance liquid chromatography (HPLC) [20-22], thin-layer chromatography (TLC) [23-34], polyacrylamide gel electrophoresis (PAGE) [35-39], capillary electrophoresis (CE) [40-46], and capillary electrochromatography (CEC) [47], for better characterization and determination of analytes.

This chapter presents results of fluorescence studies of molecules or biomolecules with Py moiety as a fluorescent chromophore. In Section 2, we introduce the underlying physics of NLN and FLN spectroscopies and the basic concepts such as homogeneous and inhomogeneous broadening, the zero-phonon line (ZPL) and phonon sideband (PSB), electron-phonon (el-ph) coupling strength, and the information these spectroscopic techniques provide. This is followed by a sampling of applications in different systems, including conformational studies and adducts identification. This chapter ends with concluding remarks and future prospects.

2. Basic Aspects of Fluorescence Line-Narrowing (FLN) Spectroscopy

FLN spectroscopy employs monochromatic laser light to selectively excite emission from chromophores within an inhomogeneously broadened ensemble of pigment-protein complexes. The resulting line-narrowed fluorescence spectrum can be related to the homogeneously broadened fluorescence spectrum. This technique has been widely used in studies of various stable and depurinating DNA adducts ([48] and references therein), chlorophyll-like molecules [49-54], bacteriochlorophylls [55, 56], and various photosynthetic complexes containing fluorescent chromophores ([57] and references therein). In this section, the fundamentals of FLNS and delta FLNS (ΔFLNS), and the information they provide, are briefly described.

2.1. Homogenous and Inhomogeneous Broadening

Generally speaking, when chemically identical chromophores experience the same environment, e.g., when embedded in a perfect crystal, these chromophores absorb exactly the same energy and give a narrow Lorentzian-shaped absorption spectra, which are homogeneously broadened [58] as shown in Figure 1A.

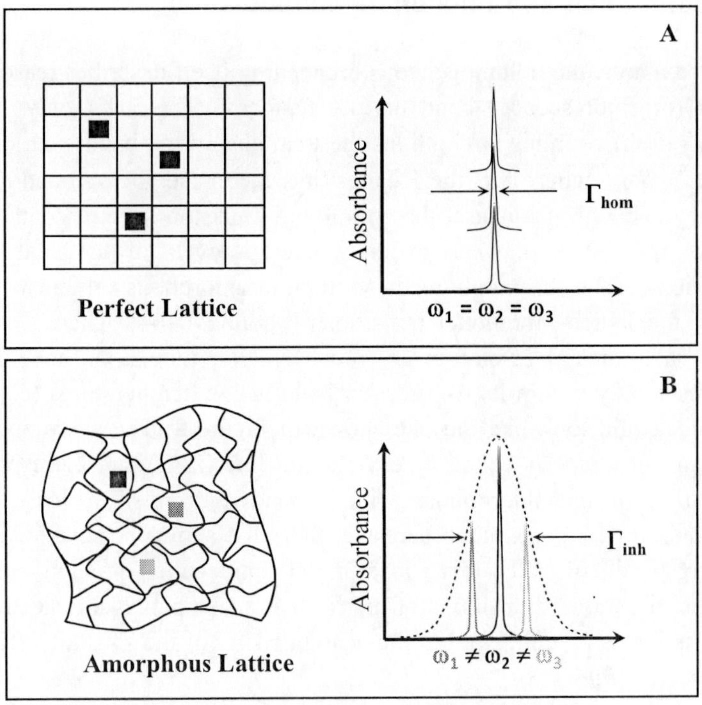

Figure 1. Schematic view of homogeneous and inhomogeneous broadening. Frame A: Chromophores are doped in highly ordered system (crystalline environment) and absorption lines are only homogenously broadened. Frame B: Molecules are in a disordered matrix so the absorption is inhomogeneously broadened.

The full-width at half maximum (FWHM) of these spectral lines can be defined as homogeneously broadened line width, Γ_{hom}, which can be directly related to the total dephasing time (T_2) of the optical transition by

$$\frac{1}{T_2} = \pi c \Gamma_{\text{hom}} = \frac{1}{2T_1} + \frac{1}{T_2'} , \tag{1}$$

where c is the speed of light, T_1 is the excited-state lifetime, and T_2' is the pure dephasing time [59, 60].

However, when the molecules are in solution or in an amorphous solid, each fluorophore's immediate microenvironment is different, resulting in modified interactions between the molecules and their surroundings. Thus, each molecule has its own well-defined absorption energy level, leading to the total absorption of all the chromophores in the system to form a broad band, referred to as an inhomogeneously broadened absorption band (Figure 1B). Its FWHM (neglecting phonons) describes the inhomogeneous broadening (Γ_{inh}) [61-63]. The total band shape reflects the property of the micro-environment of the chromophores, i.e., it describes a distribution of energetically inequivalent sites and their coupling to an environment in which fluorophores reside.

2.2. Zero-Phonon Lines and Phonon Sidebands

As mentioned above, the inhomogeneous broadening itself describes reasonably well the (0,0)-absorption (or fluorescence) band at low temperature (~5 K) only for very weak electron-phonon (el-ph) coupling strength, as the total absorption bandwidth (origin band) is described by $\Gamma_{inh} + S\omega_m$, where S is the Huang-Rhys factor (see below) and ω_m is the mean phonon frequency. Though gas-phase absorption and emission spectra contain sharp bands corresponding to transitions between different energy levels of the analyte itself, pure electronic transitions of the chromophores in solution or amorphous solid matrices are always accompanied by the lattice vibrational transitions (phonons) [64]. Theoretically, when the laser width and instrumental resolution are much smaller than Γ_{hom}, and one excites the chromophores resonantly within its (0,0)-origin band at low temperatures (e.g., at 5 K), the resulting spectrum could look like the one shown in Figure 2. The sharp and narrow band with a Lorentzian line shape is called a zero-phonon line (ZPL) and corresponds to a pure electronic transition of the fluorophore with a bandwidth of Γ_{hom} [61]. However, the experimentally measured ZPL is much narrower than that shown schematically in Figure 2. At the low-energy side of ZPL, there is a broad band called phonon sideband (PSB), characterizing the vibrational transitions of matrix that accompanies the electronic transition of the analytes. Its intensity, compared to the peak intensity of the ZPL, depends on the el-ph coupling strength (*vide infra*).

As mentioned above, the peak of PSB, ω_m, is the mean phonon frequency. The width of the ZPL, Γ_{hom} (in cm^{-1}), is determined by the lifetime T_1 of the excited electronic level, i.e., $\Gamma_{hom} = 1/2\pi c T_1$ (see Eq. (1)).

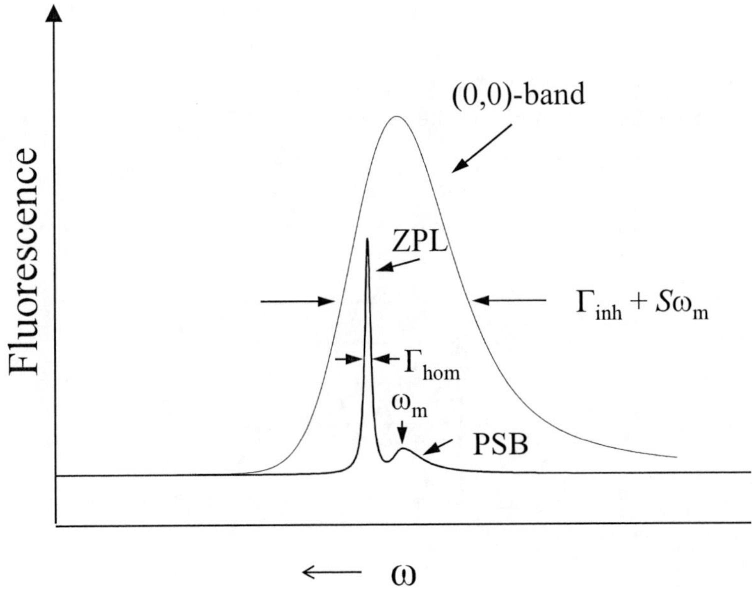

Figure 2. Schematic view of inhomogeneously broadened fluorescence origin band and fluorescence ZPL and PSB.

In absorption spectra, PSB appears at higher frequency than the corresponding ZPL; the shape of the PSB is well described as a half-Gaussian and half-Lorentzian for $\omega < \omega_m$ and $\omega > \omega_m$, respectively. In emission spectra, the PSB is located at lower energies than the corresponding ZPL, i.e., its profile is reversed on the frequency scale. From the integrated intensity of the ZPL (I_{ZPL}) and the PSB (I_{PSB}), a Debye-Waller factor α can be obtained using the following formula [12, 65]:

$$\alpha = I_{ZPL} / (I_{ZPL} + I_{PSB}). \tag{2}$$

At low-temperature limit (T ~ 0 K), the dimensionless Huang-Rhys factor is related to α by

$$S = -\ln\alpha, \tag{3}$$

and the width of the inhomogeneously broadened fluorescence origin band is also given by $\Gamma_{inh} + S\omega_m$. The physical meaning of Huang-Rhys factor is discussed in section 2.3.

2.3. Electron-Phonon Coupling Strength

This section addresses the coupling between electronic excitation of chromophore and the external lattice vibrational transition modes. Figure 3 illustrates the energy levels of a hypothetic guest chromophore in a glassy matrix. The two parabolas represent the potential energy of the ground and excited electronic states (E_0 for ground state and E_1 for excited state). Except for the lowest horizontal lines in these two states, which depict the pure electronic states of the molecule, other lines stand for the energy levels of the chromophore interacting with local phonon *i*. Note these levels are not vibrational levels of the molecule in ground or excited states. Multiple phonon quanta (energy $\hbar\omega_i$) can be described by a harmonic oscillator model (not shown in the figure).

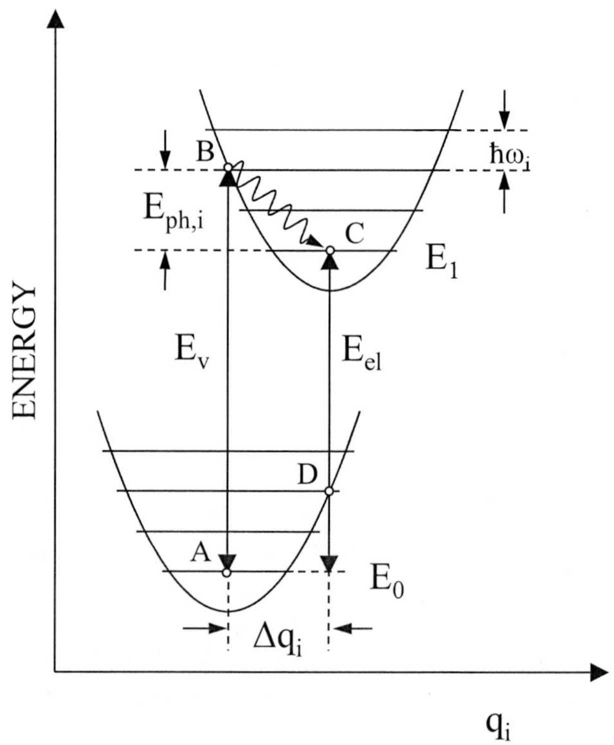

Figure 3. Configurational coordinate diagram for the analysis of the optical transition between the ground (E_0) and excited (E_1) states.

The potential energy can be expressed in the following general form [66]:

$$V = \sum_i A_i q_i + \sum_{ij} B_{ij} q_i q_j \,. \tag{4}$$

In Eq. (4), A_i is the linear coupling coefficient, describing the change of excited-state energy during optical transition because of equilibrium shift of the normal oscillator. B_{ij} *is* the quadratic coupling coefficient; for $i = j$, it describes the change in the energy of the excited state resulting from the change in the normal oscillator frequencies during an excitation. For $i \neq j$, they describe the change in the energy due to mixing of normal coordinates [18]. Our discussion below assumes a linear electron-phonon coupling approximation, where the second term of equation (4) is negligible. Here, q_i stands for the lattice normal coordinate belonging to phonon mode *i*, and Δq_i represents the change of equilibrium position. In Figure 3, E_v is the vertical transition energy, and E_{el} is the pure electronic transition energy in this process. $E_{ph,i} = E_v - E_{el}$ is the lattice vibrational transition energy involved in the vertical transition. Therefore, the strength of el-ph coupling is measured by the above-mentioned Huang-Rhys factor, S, with

$$S = E_{ph,i}/\ \hbar\omega_i. \tag{5}$$

In low-temperature limit, S is given by [12, 61, 65]

$$S(T=0) = \sum_i \frac{M_i \omega_i}{2\hbar} (\Delta q_i)^2 , \tag{6}$$

where M_i and ω_i are the reduced mass and frequency of the phonon mode i, respectively. Considering equations (2) and (3) for the relationship among S, α, and the integrated intensity of ZPL and PSB, it's obvious that when the equilibrium position shift is small from ground state to excited state, S is small, resulting in a larger α value and a larger integrated area of ZPL, as shown in Figure 4. Generally speaking, systems with $S < 1$ have weak el-ph coupling, as shown in the upper part of Figure 4. $S \sim 1$ to 3 is referred to an intermediate el-ph coupling, while $S > 3$ is observed in strongly coupled systems (lower part of Figure 4). When $S \geq 10$, the el-ph coupling is very strong for the system, and no ZPL can be observed.

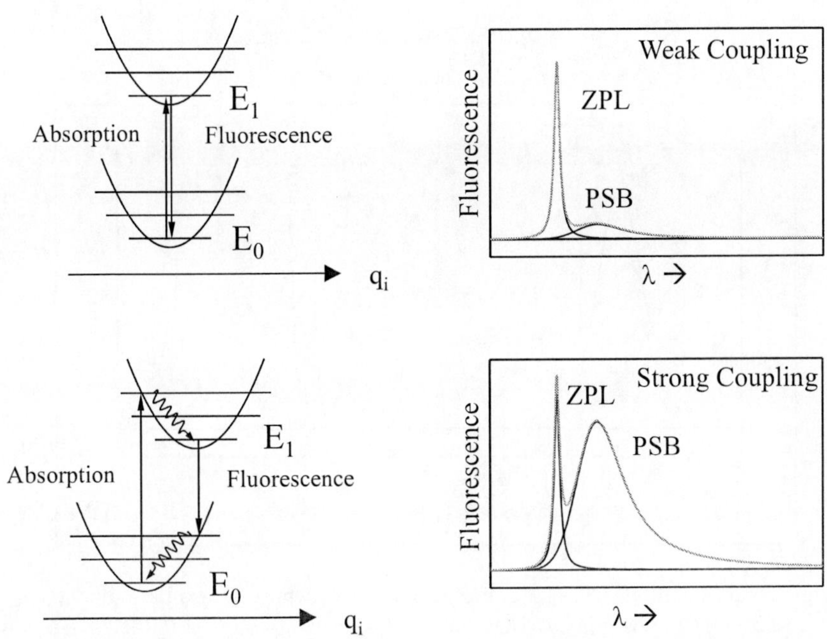

Figure 4. Diagram of potential curves in linear electron-phonon coupling and molecular line shapes of corresponding transitions.

2.4. Origin Band Excitation—Ground-State Vibrational Frequencies

When a narrow-band laser (frequency ω_L, laser linewidth $\Gamma_L << \Gamma_{hom}$) is used to selectively excite molecules within the inhomogeneously broadened (0,0)-absorption band in a glassy matrix at T ~ 5 K (origin-band excitation mode, as shown in Figure 5A), only a subset of chromophores, whose (0,0)-transition energy overlaps with that of the excitation laser, can be excited. A schematic of the resulting spectrum consisting of the ZPL (ω_{ZPL}) and two vibrational bands (with frequencies of ω_α and ω_β) is shown in the lower panel of Figure 5A. Vibrational ZPLs correspond to transitions terminating at vibrational levels of S_0. Thus, origin band excitation provides ground-state vibrational frequencies (ω_α and ω_β) and relative vibrational intensities, which, in combination, may serve as a fingerprint of studied chromophores.

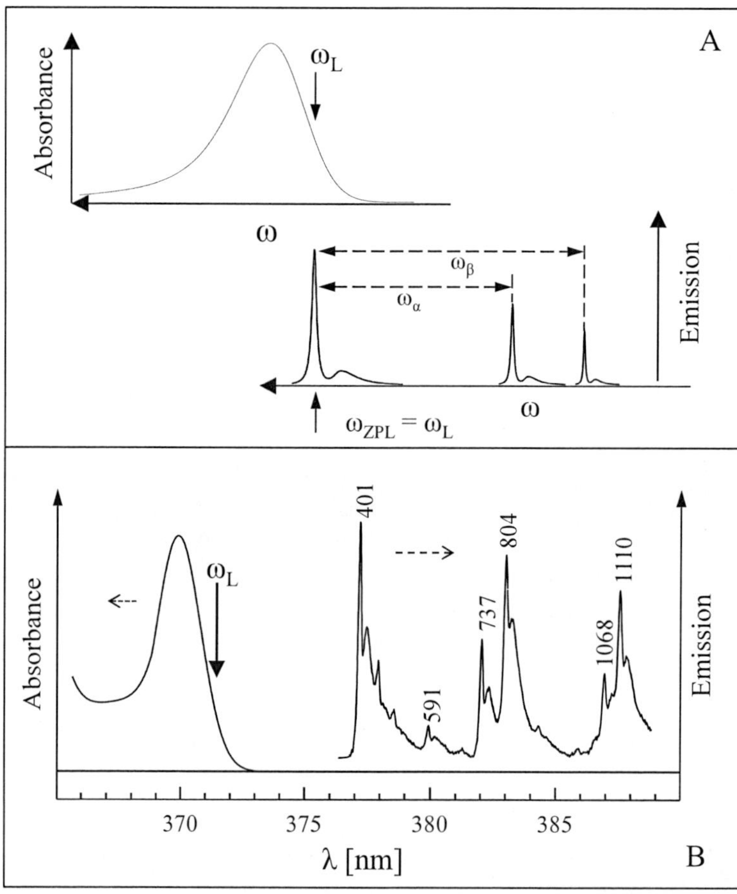

Figure 5. Origin band excitation in FLNS. Frame A: Laser frequency, ω_L, excites only those chromophores whose (0,0)-transitions overlap with the laser profile (see text for details). ω_α and ω_β correspond to the ground-state vibrational frequencies in cm^{-1}. Frame B: (0,0)-absorption band and FLN spectrum obtained for pyrene in ethanol glass at T = 4.2 K, λ_{ex} = 371.5 nm. The sharp peaks (bands) correspond to the pyrene ground-state vibrational frequencies. Note that between 373 and 376.5 nm, data were not collected.

As an example, Figure 5B shows the (0,0)-absorption band (left) and the ground-state-excited FLN spectrum of Py in ethanol glass at T =4.2 K (right).Only part of this spectrum is shown for clarity; all narrow peaks are ZPLs and the numbers above correspond to the ground-state vibrational frequencies (in cm^{-1}). Due to lattice phonon transitions coupled with electronic transitions, broad PSBs can be observed at low-energy side of ZPLs. FLN spectra with ground-state excitation can provide information of not only S_0 state vibrational modes, but also strength of el-ph coupling.

2.5. Vibronically Excited FLNS—Excited-State Vibrational Frequencies

Vibronic levels are vibrational levels of molecules in excited electronic states. Vibronically excited FLNS, which probes the S_1-state vibrational levels, is more often used for identification of closely related analytes ([10, 11, 47, 48] and references therein). In this approach, the laser is tuned to the vibrational levels in its first excited state (S_1). S_1-state vibrational dynamics, frequencies, and intensities are extremely sensitive to subtle changes in the structures of the chromophore and its environment, and this sensitivity is directly reflected in the FLN spectra obtained through this excitation mode [10, 11, 67]. Figure 6 illustrates how vibronically excited FLN spectroscopy probes S_1-state vibrational-level structure.

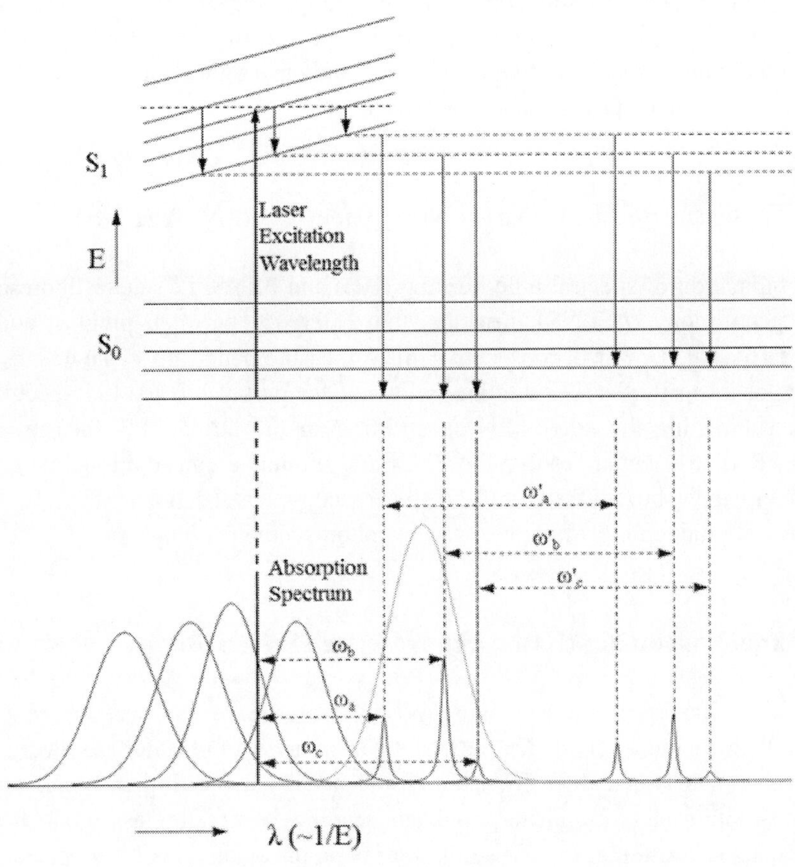

Figure 6. Jablonski diagram of FLNS and absorption cross section of vibronic transition and resulting FLN spectrum. Gray line: (0,0)-transition. El-ph coupling is neglected in this graph. Details are discussed in text.

Slanted lines in the S_1-state (upper frame) indicate (schematically) inhomogeneous broadening, corresponding to the broad absorption origin-band (gray curve) and vibronic bands represented by different Gaussian curves (lower frame). Due to inhomogeneity of the matrix, several vibronic bands may overlap on the energy scale. Therefore, laser may excite a number of overlapping vibronic energy levels. Figure 6 shows a case where three overlapping vibronic levels are excited. After relaxation to the respective zero-point levels, these three "isochromats" reach their correlated v = 0 levels of S_1 state and fluoresce. Three sharp lines within the origin-band come from the different (0,0)-transitions, and the difference between these ZPLs and laser excitation frequency gives excited-state vibrational frequencies at ω_a, ω_b, and ω_c. Furthermore, transitions also happen between v = 0 level in S_1 state and vibrational levels in ground state, resulting in less intense vibronic lines, providing information on the ground-state vibrational frequencies (ω'_a, ω'_b, and ω'_c). Of course the multiplet of ZPLs observed within the fluorescence origin band depends not only on excitation wavelengths but also on the absorption cross section of the vibronic transitions. In actual experiments, typically several ZPLs are observed that contribute to the multiplet origin structure for each excitation frequency, providing many "fingerprints" of studied molecules. Since the pattern of ZPLs varies from molecule to molecule, this spectroscopy can be used for chemical identification, conformational analysis, and/or probing the microenvironment of DNA (or protein) adducts ([10, 11] and references therein).

2.6. Delta Fluorescence Line-Narrowing Spectroscopy (ΔFLNS)

As a combination of spectral hole burning (HB) and FLNS, i.e., delta fluorescence line-narrowing spectroscopy (ΔFLNS) measures the difference between preburn and postburn FLN spectra [68-71], i.e., two consecutive FLN measurements are separated by a certain period of exposure to the laser excitation, allowing for partial HB. ΔFLN spectrum is then obtained by subtracting the latter FLN spectrum from the initial FLN measurement. This method can be used to better resolve the ZPLs and minimize contributions from molecules that absorb via their broad PSB. ΔFLN spectroscopy has been successfully applied to investigate the el-ph coupling strength in various photosynthetic complexes [54, 72-77].

2.7. Non-Line-Narrowing (NLN) Fluorescence Spectroscopy

In the conventional non-line-narrowing (NLN) fluorescence spectroscopy, molecules can be excited to their second excited electronic state or higher energy states. As a result, due to a lack of correlation between different excited states, all sites within the inhomogeneously broadened absorption band might be populated, leading to a featureless broad fluorescence spectrum. Such spectra, though very useful, are not as informative as FLN spectra discussed above. Nevertheless, NLN spectra can still provide important information such as the maximum of the (0,0)-transitions and the FWHM of the emission band, which are useful for probing the interaction between a chromophore and its environment. Also, NLN spectra are prerequisites for FLN spectra collection, as they determine the detection range and excitation wavelengths for both the origin-band and vibronically excited FLN spectra.

3. Applications of FLN Spectroscopy to Molecules with Pyrene Moiety

3.1. Py in Glassy Matrices

As mentioned above, Figure 5B shows a fraction of the FLN spectrum of Py in ethanol glass with ~400-1100 cm^{-1} ZPLs; the most prominent ground-state vibrational modes observed for Py are listed in Table I with the modes assignment based on ref. [78].

Table I. Comparison of ground-state vibrational modes of Py obtained by FLNS, Raman spectroscopy, and calculation

FLN vibrational modes	Raman*	Calc.*	Assignment*
401	408	401	a_g
591	591	586	a_g
737	733	746	b_{3g}
804	805	791	a_g
1068	1067	1064	a_g
1110	1107	1116	b_{3g}

* From ref. [78].

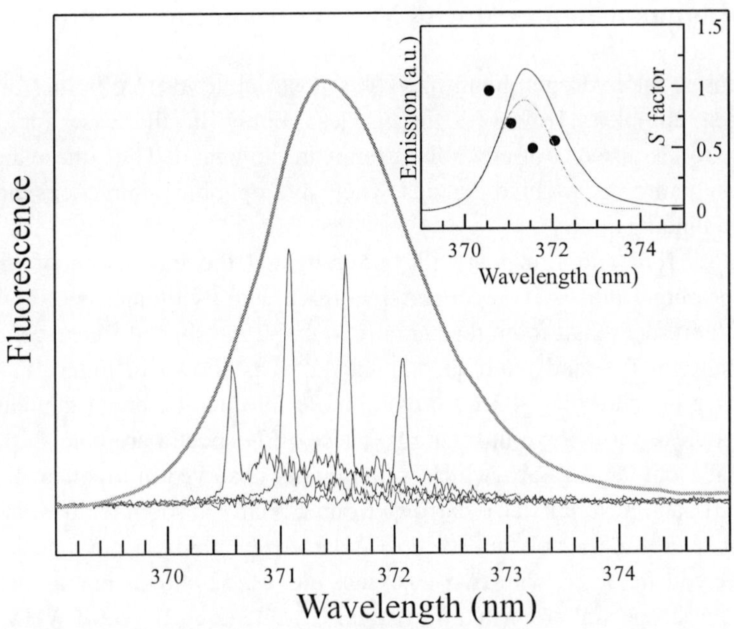

Figure 7. Fluorescence origin-band (thick gray line) and origin-band excited ΔFLN spectra (black curves) of Py in ethanol glass at T = 5 K obtained at various excitation wavelengths (from left to right: 370.58 nm, 371.08 nm, 371.58 nm, and 372.08 nm) using time-resolved measurements. Gate width and delay time are 200 ns and 40 ns, respectively. The inset compares the (0,0)-emission band (black) with a Gaussian fit of the side distribution function (SDF) shown as a gray spectrum. Frequency-dependent *S* factor is plotted as a function of excitation wavelength (see filled black circles) [79].

From Table I, it is obvious the S_0-state vibrational frequencies collected by the FLNS method are in good agreement with data obtained through Raman spectroscopy and theoretical calculation [78]. Figure 7 shows fluorescence origin band (thick gray line) and origin-band excited ΔFLN spectra (black curves) of Py in ethanol glass at T = 5 K obtained for four laser excitations within the origin band (see figure caption for details). To ensure that scattered- light intensity does not contribute to the resonant ZPLs, the spectra were obtained using time resolved measurements. The inset in Figure 7 compares the (0,0)-emission band (black spectrum) with a site-distribution function (SDF; gray curve) obtained by a Gaussian fit of constant-fluence ZPLs. Frequency-dependent S factors are plotted as a function of excitation wavelength (see filled black circles) [79]. The dependence of coupling strength on excitation wavelengths is discussed in detail elsewhere [80, 81].

Regarding various NLN spectra for pyrene in various solutions, Nakajima [82] and Kalyanasundaram *et al.* [83] reported in the 1970s that not only can electronic transitions be affected by solvent effects, but vibrational bands of electronic spectra also vary depending on matrix composition. Generally speaking, when the solvent is less polar, (0,0)-transitions of the solute with Py chromophore will be red-shifted and the vibronic bands will more intense when compared with the (0,0)-bands. This property is very important in the structural and conformational studies for B[*a*]P metabolites as BPT and BPDE-DNA adducts with pyrenyl moiety [84, 85], which are discussed in sections 3.3-3.6.

3.2. Pyrene Immunocomplexed with Various Monoclonal Antibodies (mAbs)

In many cases, the hydrogen bond plays the most important role in the formation of an antibody-antigen complex. However, this is less likely in the case for PAH-binding antibodies, due to the absence of any polar groups in the ligand. Thus, these antibodies may follow different pathways to bind ligands when hydrophobic interactions dominate their recognition and binding ability.

Grubor *et al.* [86] and Lin *et al.* [87] investigated the interaction of *anti*-polycyclic aromatic hydrocarbon (*anti*-PAH) mAb and 4D5 mAb with Py by means of low-temperature laser-induced fluorescence. At room temperature, only ~2 nm shift of fluorescence peak could be observed between free and immunocomplexed Py, as shown in Figure 8. Fluorescence spectra at low temperature (T = 4.2 K) provide more information about the conformation of the immunocomplexes, as shown in Figure 9. These NLN spectra are blue-shifted relative to data collected at room temperature, which is commonly observed in low-temperature spectra. FWHM for (0,0)-band of free Py also narrows from 235 cm^{-1} at room temperature to 145 cm^{-1} at 4.2 K. Most interestingly, for the Py/*anti*-PAH mAb complex, a clear peak-splitting was observed that could be fitted with two Gaussian curves, as shown in Figure 9A. Spectral properties of these spectra are listed in Table II. Due to different FWHM of the two contributions deconvoluted from spectra for the Py/*anti*-PAH mAb complex, *anti*-PAH mAb appears to bind Py by adopting two distinct conformations of its binding sites. FLN spectra were also collected for Py in various solvents to investigate the interaction between mAb and Py, as shown in Figure 10. Glycerol/PBS (1:1) matrix and ethanol were used as protic solvents, methylcyclohexane as an aprotic, aliphatic solvent, and dimethyl sulfoxide and toluene as π-electron-containing solvents. The spectrum for immunocomplexed Py is most similar to curve e, i.e., the spectrum of Py in toluene, which indicates that π-π interactions are most likely responsible for binding Py in *anti*-PAH mAb.

Table II. Spectral characteristics of Py/*anti*-PAH mAb and Py/4D5 mAb complexes in PBS and G/PBS (1:1) matrices at 4.2 K[a].

Matrix	Pyrene/anti-PAH mAb complex				Pyrene/4D5 mAb complex				Pyrene	
	Conformation I[b]		Conformation II[b]		Conformation I		Conformation II			
	(0,0)[c] ±0.2 nm	FWHM ±5 cm^{-1}	(0,0) ±0.2 nm	FWHM ±5 cm^{-1}	(0,0) ±0.2 nm	FWHM ±5 cm^{-1}	(0,0) ±0.2 nm	FWHM ±5 cm^{-1}	(0,0) ±0.2 nm	FWHM ±5 cm^{-1}
PBS buffer	372.9	75	374.4	125	372.3	75	—	—	371.6	160
G/PBS Matrix	372.4	75	373.6	125	372.1	75	—	—	371.1	160

[a] Reprinted with permission from ref. [87].
[b] Conformation I and II correspond to the conformations fitted with dotted and dashed Gaussian curves, respectively.
[c] Maximum of fluorescence (0,0)-band (in nm).

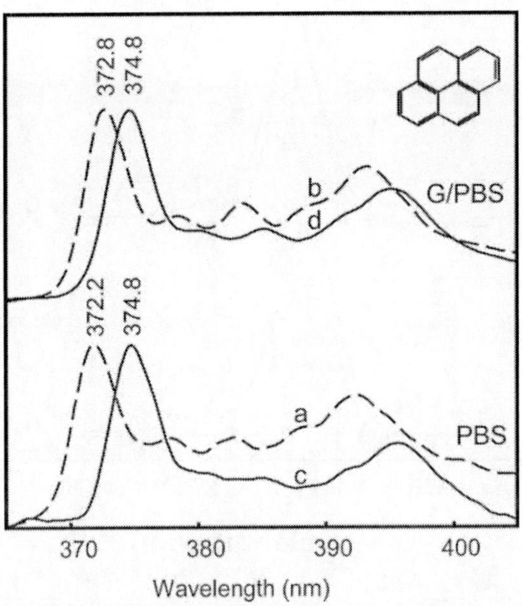

Figure 8. Normalized fluorescence emission spectra of free Py and Py/mAb complexes in PBS buffer (bottom) and G/PBS matrix (top) at room temperature: a, free Py in PBS buffer; b, free Py in G/PBS matrix; c, Py/mAb complex in PBS buffer; d, Py/mAb complex in G/PBS matrix. Reprinted with permission from [86]. Copyright (2005) National Academy of Sciences, U.S.A..

The nature of band I in the NLN spectra for the Py/*anti*-PAH mAb complex was further investigated in ref. [87]. Comparison of frames A and B in Figure 9 indicates that 4D5 mAb binds Py in a single conformation of its binding site. Interestingly, its origin-band position and FWHM are similar to those obtained for conformation I of the Py/*anti*-PAH mAb complex. This indicates that when these two different mAbs adopt conformation I to bind Py, they provide the chromophores with similar inhomogeneous broadening. The binding-site structure of 4D5 mAb is known from extensive modeling studies [88, 89], as shown in Figure 11.

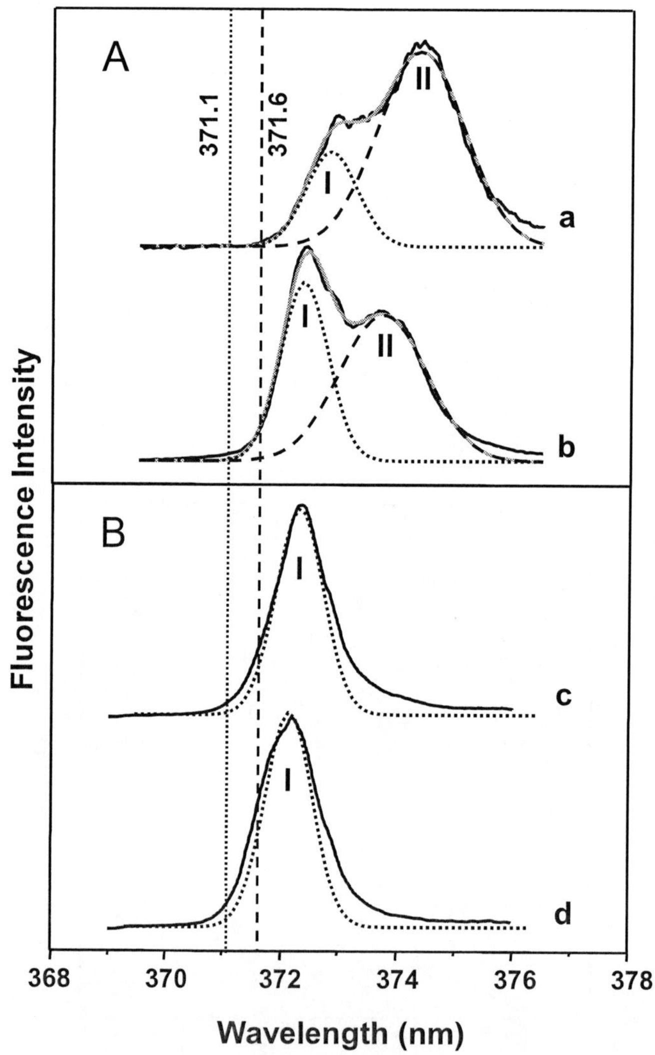

Figure 9. Low-temperature NLN fluorescence spectra of immunocomplexed pyrene (Py) in two different mAbs, i.e., *anti*-PAH (A) and 4D5 (B), respectively. T = 4.2 K; gate width, 200 ns; delay time, 40 ns; λ_{ex} = 308 nm. Spectra a and c represent the immunocomplexed Py in PBS buffer, and spectra b and d in glycerol/PBS buffer mixture (1/1). The dotted and dashed vertical lines indicate the position of the (0,0)-band of free Py in PBS buffer and Glycerol/PBS buffer mixture, respectively. The bands assigned to conformations I and II are indicated by the dotted and dashed Gaussian lines, respectively. Reprinted with permission from [87].

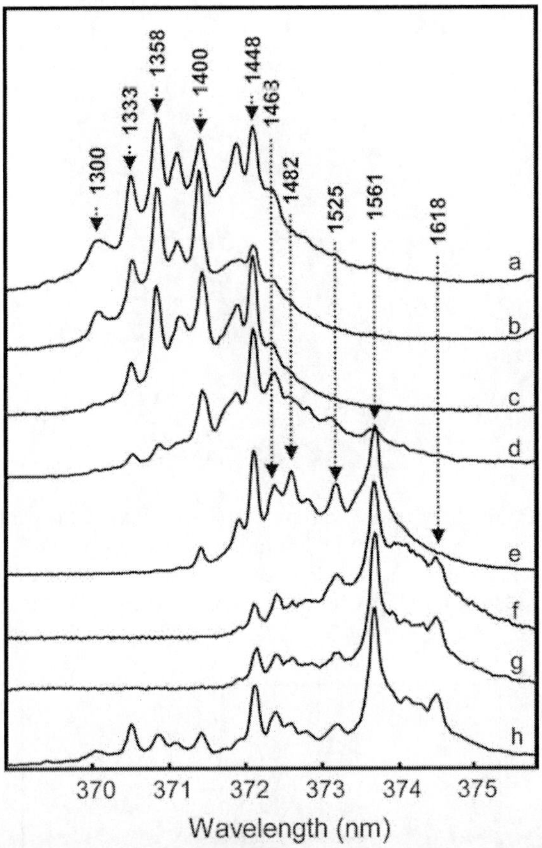

Figure 10. FLN spectra of Py in various solvents and Py/mAb complex in two matrixes: a, G/PBS; b, methylcyclohexane; c, ethanol; d, dimethyl sulfoxide; e, toluene; f, Py/mAb complex in PBS; g, Py/mAb complex in G/PBS; h, Py/mAb complex with an excess of Py in G/PBS. $\lambda_{ex} = 353$ nm; delay, 40 ns. Numbers in bold represent excited-state vibrational energies in cm^{-1}. Reprinted with permission from [86]. Copyright (2005) National Academy of Sciences, U.S.A..

It was suggested ArgH95 inside the binding pocket of 4D5 mAb be critical in hapten binding, as ELISA showed the affinity between 4D5 mAb and various PAH ligands dramatically decreased in the ArgH95Gln mutant [89]. It appears that instead of a regular π-π interaction, this positively charged residue uses a π-cation interaction to stabilize a neutral aromatic ligand [88, 89]. Considering the similar inhomogeneity observed in NLN spectra for Py/4D5 mAb and conformation I of Py/*anti*-PAH mAb, it is reasonable to suggest that *anti*-PAH mAb may also use positively charged residues to stabilize Py in conformation I of the mAb binding site. The latter cannot be confirmed by modeling studies, as the structure of the *anti*-PAH mAb is not known. This conclusion is also supported by FLN spectra shown in Figure 12. Note that spectra a-c are distinctly different from each other due to different band maxima and different Γ_{inh}. These spectral changes are induced by different immediate molecular environments of Py molecules. Because of the contribution from two conformations for Py/*anti*-PAH mAb, I and II (*vide supra*), spectrum c in Figure 12 shows vibrations over a much broader energy range (due to broader origin bands) than spectrum b. Note that free Py in a glassy matrix (curve a) also reveals more vibrational modes within its (0,0)-band due to a larger inhomogeneous broadening ($\Gamma_{inh} \sim 160$ cm^{-1}).

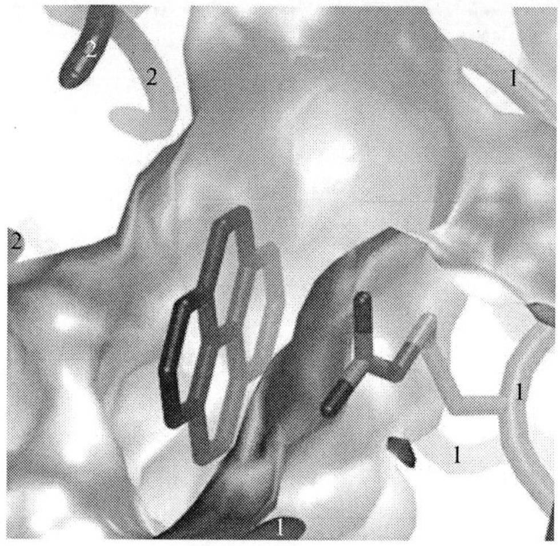

Figure 11. Model of the 4D5 mAb fragment with pyrene (dark sticks in the center) docked in the lowest binding energy conformation in the huge antibody-binding pocket represented by its molecular surface. The tube labeled 1 is the heavy chain; the tube labeled 2 corresponds to the light chain. Only the ArgH95 and pyrene are displayed for clarity. Graphics was made using VMD. Reprinted with permission from [87].

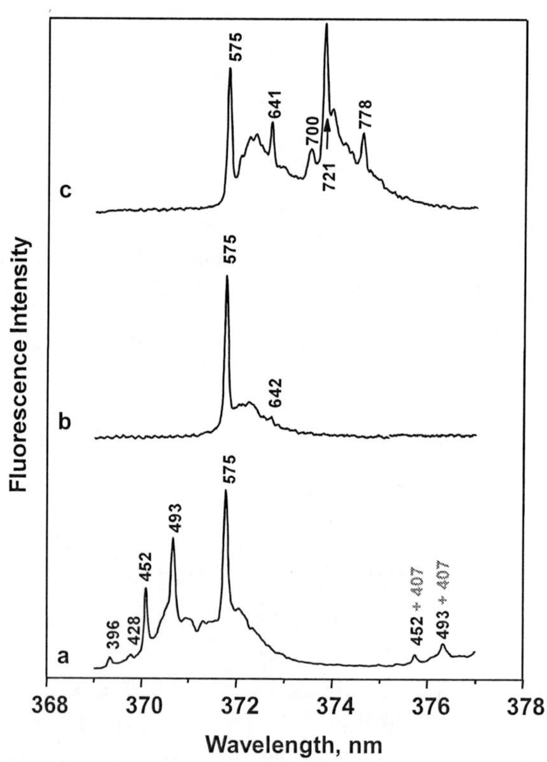

Figure 12. FLN spectra of Py in various mAbs at T = 4.2 K obtained with an excitation wavelength (λ_{ex}) of 364 nm (gate width = 200 ns, delay time = 40 ns). Spectra a – c correspond to free Py, Py/4D5 mAb complex, and Py/*anti*-PAH complex in glycerol/PBS (1/1), respectively. In spectrum a, the 407 cm^{-1} mode corresponds to the ground state vibrational frequency of Py building on the zero-phonon lines at 452 and 493 cm^{-1}, respectively. Reprinted with permission from [87].

Comparison of curves b and d in Figure 9 with spectra c and b of Figure 12 (for λ_{ex} = 364 nm) clearly indicates that the 575 cm^{-1} mode observed in curve c in Figure 12 corresponds to Py/*anti*-PAH mAb in conformation I. Higher frequency modes at 641 cm^{-1} (at least in part) and at 700, 721, and 778 cm^{-1} correspond to red-shifted conformation II. Finally, authors of ref. [87] argued that the nature of blue-shifted conformation with a very narrow fluorescence (0,0)-band (FWHM ~75 cm^{-1}) is consistent with π-cation interactions, underscoring the importance of π-cation interaction in ligand binding and stabilization in agreement with modeling studies. It has been suggested that considerable narrowing of the fluorescence origin-band of ligand in the protein environment could be regarded as a simple indicator of π-cation interactions.

3.3. Identification of Enantiomers of Benzo[*a*]Pyrene Tetraols

Benzo[*a*]pyrene (B[*a*]P) is a well-studied carcinogen, which forms stable DNA adducts through its bay-region diolepoxide (BPDE) intermediate [90-92] and dupurinating adducts via a one-electron oxidation mechanism [93]. B[*a*]P-tetraols (BPTs) are also of interest, as they are the final hydrolytic products of BPDE-derived DNA adducts. However, compared to B[*a*]P, fluorescence properties of BPTs are closer to those of Py due to the oxidation of the α benzene ring (ring A in Figure 13). Thus, BPTs have Py moiety as a fluorescent-functional group.

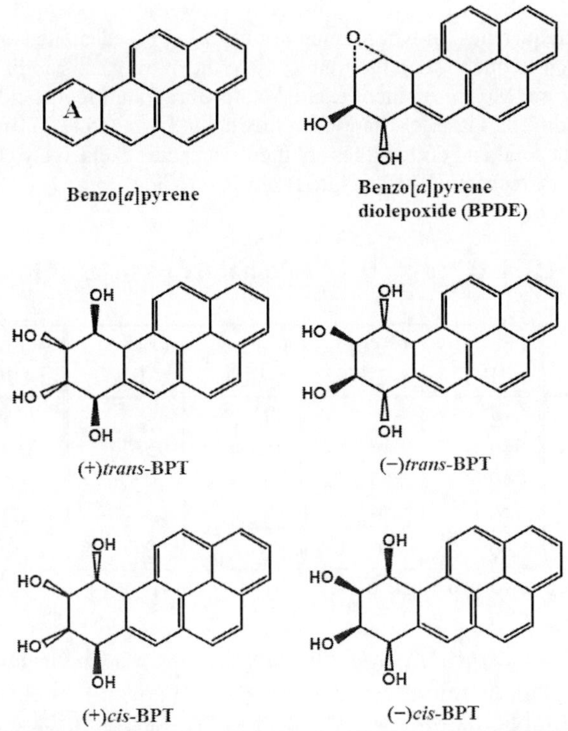

Figure 13. Molecular structures of B[a]P, BPDE, (±)cis-anti-BPT, and (±)trans-anti-BPT.

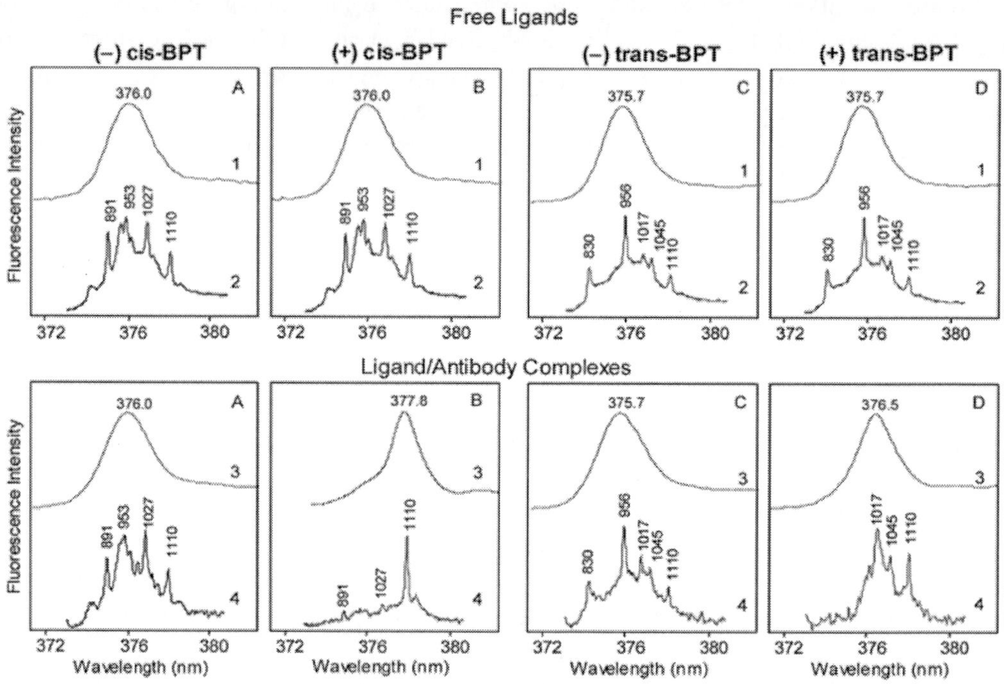

Figure 14. Frames A–D refer to (–)*cis*-BPT, (+)*cis*-BPT, (–)*trans*-BPT, and (+)*trans*- BPT, respectively. T = 4.2 K, gate width = 200 ns, gate delay = 40 ns. NLN/FLN spectra of free and immunocomplexed BPT are labeled 1/2 and 3/4, respectively. Vertically oriented numbers label excited-state vibrational frequencies (in wavenumbers; $\Delta\nu = \pm 3\ cm^{-1}$) calculated as the differences between the laser excitation frequency and the frequency of the narrow emission line in the corresponding FLN spectrum. Most prominent excited-state vibrational frequencies of *cis*-BPT and *trans*-BPT obtained with different laser excitations in the region from 355 to 371 nm are listed in Table III. Values underlined with solid and dotted lines are the frequencies exclusively characteristic for cis-BPT and trans-BPT isomers, respectively. Reprinted with permission from [94]. Copyright 2006 American Chemical Society.

Table III. Excited-state vibrational frequencies of BPT*

(+)/(–) c-BPT	(+)/(–) t- BPT	(+)/(–) c-BPT	(+)/(–) t- BPT	(+)/(–) c-BPT	(+)/(–) t- BPT	(+)/(–) c-BPT	(+)/(–) t- BPT
453	449	617	–	891	–	1110	1110
479	–	646	646	953	956	1384	1383
495	503	740	715	–	1017	1445	1440
538	546	770	760	1027	–	1562	1562
578	578	830	830	–	1045		

* For details see figure caption of Figure 14.

As shown in Figure 13, BPT has four stereoisomers, which are very difficult to differentiate by low-resolution fluorescence methods (see spectra labeled 1 in all frames of Figure 14) [94]. For example, only a 0.3-nm shift of (0,0)-band is observed between *cis*- and *trans*-enantiomers. Although FLN spectra can easily differentiate between *cis*- and *trans*-isomers, they still fail to reveal any difference between (+)- and (–)-enantiomers (see curves labeled as 2 in the top panels for free ligands in Figure 14). It is well-established that chiral differentiation requires a chiral element in the analytical method. Thus, *anti*-PAH mAb can be injected into these systems (during the CE separation process) to form complexes with the four isomers. As shown in the lower panels in Figure 14, fluorescence spectra of immunocomplexed (–)-enantiomers didn't experience much change. However, (0,0)-positions of (+)-BPT/mAb complexes are more red-shifted, and relative intensities of ZPLs in FLN spectra are dramatically different from those observed for the free ligands (top panels), due to different shifts of the respective origin bands. Recall that *anti*-PAH mAb (see section 3.2) can bind pyrene via both π-π and π-cation interactions. Therefore, *anti*-PAH mAb could use the same interactions to bind BPT, although additional steric interactions must be also important due to the presence of four hydroxyl groups. The latter could advance or hinder a molecule's favorable orientation and interaction with the mAb-binding site. Figure 15 shows the determination of association constants (K_a) for four BPT isomers immunocomplexed with *anti*-PAH mAb to correlate their spectral behavior and binding strength. It was shown that (+)-isomers are more favorable in the mAb-binding site than the (–)-isomers, which is in accordance with the lager red-shift of spectra observed for the (+)-BPT/mAb complexes.

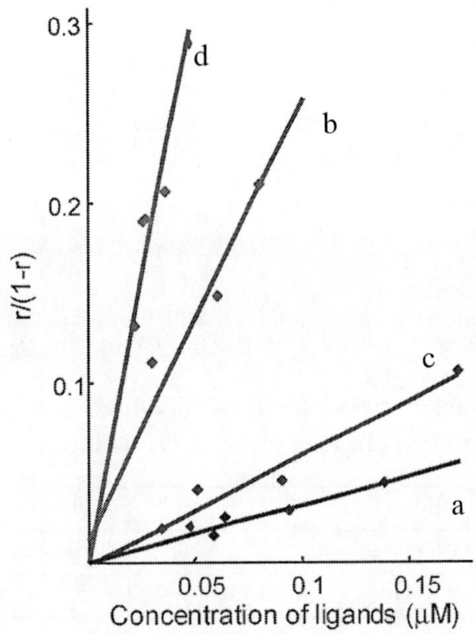

Figure 15. Plot of $r/(1-r)$ against the concentration of free ligands, where r is the fraction of ligand bound per mAb. The association constants (K_a's) for various BPT/*anti*-PAH MAb complexes are derived from the slopes of the linear regression lines: (+)-*trans*-BPT, line d, $K_a = 6.0 \times 10^6$ M^{-1}; (+)-*cis*-BPT, line b, $K_a = 2.4 \times 10^6$ M^{-1}; (–)-*trans*-BPT, line c, $K_a = 6.1 \times 10^5$ M^{-1}; (–)-*cis*-BPT, line a, $K_a = 2.1 \times 10^5$ M^{-1}. Reprinted with permission from [94]. Copyright 2006 American Chemical Society.

3.4. Identification of DNA Adducts Using Monoclonal Antibodies

The methodology discussed briefly above, to identify immunocomplexed enantiomers by FLN spectroscopy, can be extended to differentiate various BPDE-derived DNA adducts shown in Figure 16 [95]. Although the latter could not be accomplished by using *anti*-PAH mAb (raised against fluoranthene), which has too small binding pocket to adopt bulky *anti*-BPDE deoxyguanosine (dG) adducts, another monoclonal antibody, i.e., 8E11 mAb (raised against B[*a*]P-DNA conjugates), was successfully used in DNA adduct identification and/or differentiation [95].

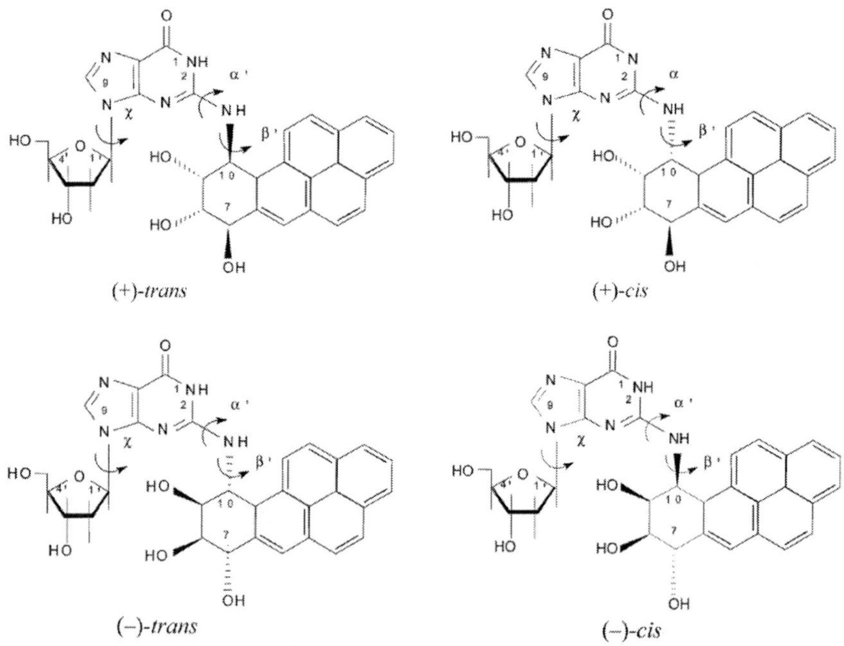

Figure 16. Structures of (±)-*anti-trans*- and (±)-*anti-cis*-BPDE–N^2-dG adducts. Torsion angles χ, α', and $\beta'$are defined as follows: χ, O4′–C1′ –N9–C4; α', N1–C2–N2–C10 (BP); and β', C2–N2–C10 (BP).

Table IV. Spectral characteristics of BPDE–dG adducts and BPDE–dG adduct/8E11 mAb complexes in PBS/glycerol (1:1) matrix at 4.2 K*

	Adduct		Adduct/8E11 mAb complex	
BPDE-dG Adducts	$(0,0)^a \pm 0.2$ nm	$\Gamma^b \pm 5$ cm^{-1}	$(0,0)^a \pm 0.2$ nm	$\Gamma^b \pm 5$ cm^{-1}
(+)-*anti-trans*	376.1	140	376.7	70
(−)-*anti-trans*	376.1	140	377.2	80
(+)-*anti-cis*	376.4	145	377.8	125
(−)-*anti-cis*	376.4	145	377.1	115

[a] The maximum of fluorescence (0,0)-origin band (in nm).

[b] Γ is the FWHM obtained by doubling the half-width at half maximum measured at the high energy side of the band (in cm^{-1}).

* Reprinted with permission from [95]. Copyright 2007 American Chemical Society.

The properties of NLN fluorescence spectra for free and immunocomplexed adducts are listed in Table IV, and corresponding FLN spectra are shown in Figure 17. Smaller number of peaks (for the same laser-excitation wavelength) indicates a smaller inhomogeneous broadening, i.e., smaller disorder of the BPDE-dG adducts inside the binding site of 8E11 mAb. This is in agreement with the affinity study [95], which shows that *trans-anti*-BPDE-dG adducts, with much narrower (0,0)-band, have a higher affinity toward 8E11 mAb, i.e., they could bind in a well-defined conformation. FLN spectra shown in Figure 17 show different spectral patterns for all adduct/mAb complexes proving that all four stereoisomers of *anti*-BPDE-dG adducts can be easily differentiated. Although in frame A the difference between curves b and c is not as clear as that between the same pairs of spectra in other frames, it is still easy to identify immunocomplexed (+)-*trans*-BPDE–dG and (−)-*trans*-BPDE–dG by applying another excitation wavelength on the system, as shown in frame C. On the other hand, dramatic differences among all three spectra in frame B, in particular the changes in vibronic modes, indicate that within the mAb binding pocket, different conformational domains (and/or domain mixtures) are possible. Finally, spectroscopic study of Py/B[*a*]P immunocomplexed with 8E11 [87] suggests the possibility this mAb can use π-cation interaction to bind PAH molecules. Considering smaller inhomogeneity associated with π-cation interactions (see section 3.2), the authors of ref. [87] concluded that at least *trans*-BPDE–dG adducts may be stabilized by π-cation interaction within the binding site of 8E11 mAb.

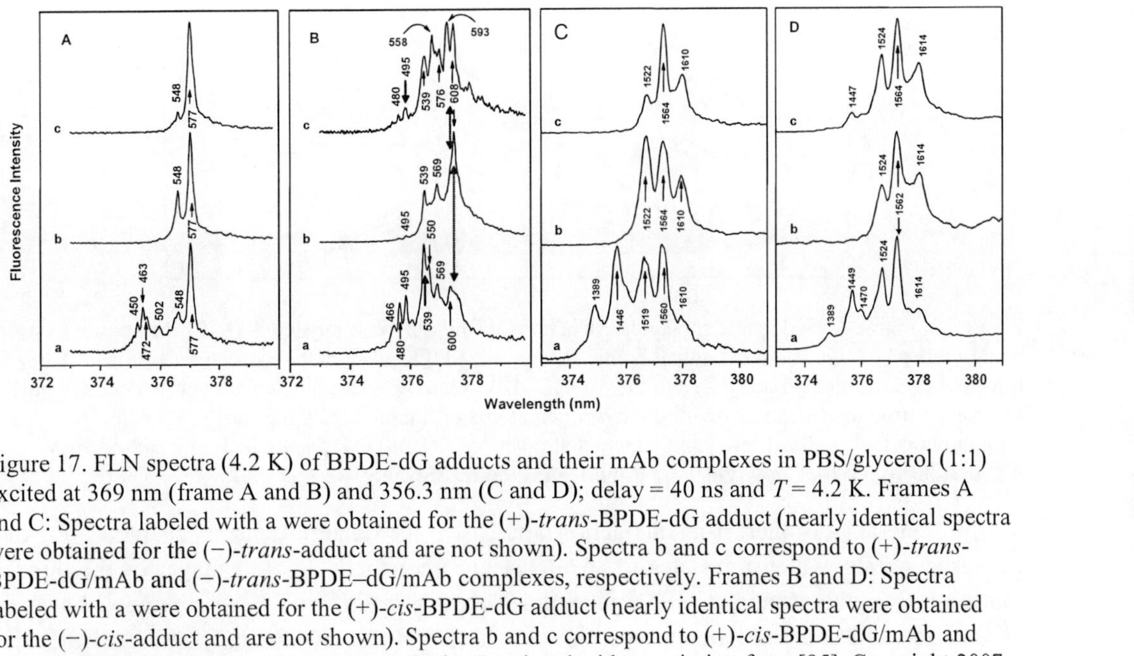

Figure 17. FLN spectra (4.2 K) of BPDE-dG adducts and their mAb complexes in PBS/glycerol (1:1) excited at 369 nm (frame A and B) and 356.3 nm (C and D); delay = 40 ns and $T = 4.2$ K. Frames A and C: Spectra labeled with a were obtained for the (+)-*trans*-BPDE-dG adduct (nearly identical spectra were obtained for the (−)-*trans*-adduct and are not shown). Spectra b and c correspond to (+)-*trans*-BPDE-dG/mAb and (−)-*trans*-BPDE–dG/mAb complexes, respectively. Frames B and D: Spectra labeled with a were obtained for the (+)-*cis*-BPDE-dG adduct (nearly identical spectra were obtained for the (−)-*cis*-adduct and are not shown). Spectra b and c correspond to (+)-*cis*-BPDE-dG/mAb and (−)-*cis*-BPDE-dG/mAb complexes, respectively. Reprinted with permission from [95]. Copyright 2007 American Chemical Society.

3.5. Py-Labeled Catechol Estrogen-Derived Metabolites and DNA Adducts

Because of the limitation of conventional absorption and electrochemical methods to identify (non-fluorescent) estrogen metabolites and estrogen-DNA adducts, 4-hydroxy-catechol estrogens and its DNA depurinating adducts (i.e., the 4-OHE$_1$(E$_2$)-1-N^7-Gua adduct standards and 4-OHE$_1$(E$_2$) metabolites) have been derivatized with 1-pyrenesulfonyl chloride and investigated by low-temperature fluorescence spectroscopy under non-line narrowing

(NLN) and FLN conditions [96]. Figure 18 provides the schematic and optimized structure for dipyrene-labeled 4-OHE_2 metabolite and 4-OHE_2-1-N^7-Gua adduct [96]. Introduction of two pyrenyl moieties in one molecule greatly increases intramolecular π-π interaction—in 4-OHE_2-1β-N^7-Gua-dipyrene adduct, Py moiety 2 adopts a stacked conformation with the base, while Py moiety 1 is nearly parallel to the estrogen A ring (Figure 18B right); in 4-OHE_2-dipyrene metabolite, both Py moieties are parallel to the aromatic ring in estrogen (Figure 18B left).

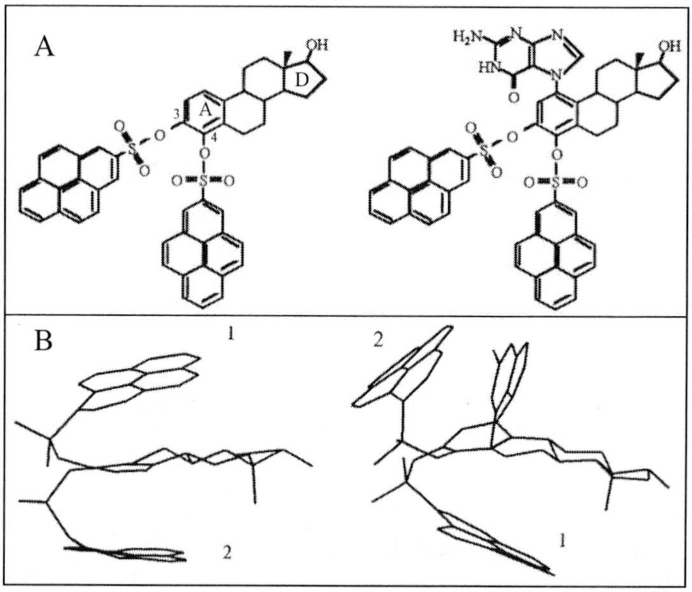

Figure 18. Frame A: Schematic molecular structures of the dipyrene-labeled 4-OHE_2 metabolite (left) and 4-OHE_2-1-N^7-Gua adduct (right). Frame B: Optimized 0 K ground-state structures of the 4-OHE_2-dipyrene-labeled metabolite (left) and the 4-OHE_2-1-N^7-Gua-dipyrene-labeled adduct (isomer β, right). Molecular structures are shown without hydrogen atoms or double bonds for clarity. Pyrene chromophores labeled as 1 and 2 are attached at positions O3 and O4, respectively. Reprinted with permission from [96]. Copyright 1998 American Chemical Society.

In general, stronger π-π interaction increases the charge-transfer character of the fluorescent state, which translates to an increasing permanent dipole change for the optical transition from the ground state. Larger dipole moment leads to increasing geometry change of the immediate environmental matrix, i.e., larger electron-phonon coupling. Thus, the measured FLN spectra contain weak ZPLs and strong PSBs, as shown in Figure 19. Compared to curves a and b, curves c and d have no resolved ZPLs, indicating stronger π-π interaction in the 4-OHE_2-1-N^7-Gua-dipyrene adduct than in 4-OHE_2-dipyrene metabolite, possibly due to the short distance between Py moiety and the guanine base.

These standards and their FLN spectra have subsequently been used for the spectroscopic identification of depurinating DNA adducts formed in a tissue culture experiment where rat mammary gland tissue was treated with the estrogen quinone (E_2-3,4-Q). Using a fluorescently labeled catechol estrogen-3,4-quinone derived N^7-Gua adduct standard, it has been demonstrated (data not shown) that the depurinating adduct formed in rat mammary tissue was 4-OHE_2-1- N^7-Gua [96].

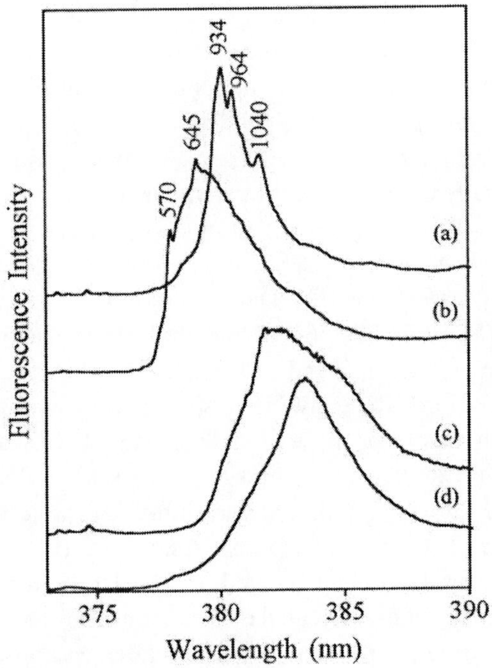

Figure 19. Vibronically excited FLN spectra obtained at 4.2 K for the 4-OHE$_2$–dipyrene metabolite (spectra a and b) and the 4-OHE$_2$-1- N^7-Gua-dipyrene adduct (spectra c and d) in ethanol. Spectra a, c and b, d were obtained with λ_{ex} of 367.0 and 370.0 nm, respectively. The FLN peaks are labeled with their excited-state vibrational frequencies in cm^{-1}. Reprinted with permission from [96]. Copyright 1998 American Chemical Society.

3.6. Structural and Conformational Studies of BPDE-DNA Adducts and BPT/DNA Physical Complexes

It is important to characterize and determine the nucleotide, nucleoside, and macro-molecular DNA adducts to understand the initiating steps in chemical carcinogenesis. Low-temperature laser-induced fluorescence has been applied in these studies for more than 20 years ([48] and references therein).

For BPDE-DNA adducts, there are two major conformations termed type I (pyrenyl moiety nearly parallel to the planes of unmodified base, referred to as interior, intercalated, or quasi-intercalated sites) or II (lower degree of intercalation between aromatic Py plane and bases, referred to as exterior sites) [84, 97]. For adduct characterization, the NLN spectra are of great importance, as not only the position and FWHM of (0,0)-band, but also the magnitude of the so-called R ratio can help to classify different structures and conformations of BPDE-DNA adducts and BPT/DNA physical complexes. In these systems [84, 85], R ratio is defined as the magnitude of the relative intensity of the fluorescence origin band over the vibronic band ~1400 cm^{-1} lower in energy, and this value decreases with a less-polar microenvironment of fluorophore(s). References [84, 97] reported the structure of adducts formed by treating calf thymus DNA with *anti*-BPDE. Table I in ref. [85] lists three and five different DNA adducts derived from (+)-*trans*-7,8-dihydroxy-*anti*-9,10-epoxy-7,8,9,10-tetrahydro-benzo[*a*]pyrene ((+)-*anti*-BPDE) and its (–)-enantiomer, respectively. Two methods, FLNS with fluorescence quenching, and R-ratio determination of different conformations for BPT/DNA complexes corresponding to different BPDE-DNA adducts, help to assign these structures to type I or II. The same group further investigated the adducts' structure by treating various polynucleotides, such as poly(dG-dC)·poly(dG-dC), poly(dG)·poly(dG-dC), single-strand poly(dG), etc., or GMP, with (+)-*anti*-BPDE. In this study [98], a new site-type feature was identified and named type III, which was assigned to the minor adenine adducts. It was suggested that in the type III adduct, the pyrenyl reside (in the vicinity of the negatively charged phosphodiester group) is responsible for high R-value.

As shown in ref. [85], spectra for BPT/DNA share some similarities with the spectra of BPDE-DNA adducts. In ref. [98], the 376.1~376.8 nm band was assigned to "free" BPT and 381.1 nm band to BPT strongly intercalated between bases, while the latter conformation still had sufficient solvent exposure. Recently NLN spectra were collected from enantiomeric *anti*-BPT complexes with native ds p53exon5 DNA and C-5 analogs containing the 5-methylcytosine (^{Me}C) or dPER [99]. Compared to native DNA, all spectra for ds p53exon5 containing ^{Me}C:G or dPER:G base pair are red-shifted by at least 1 nm, suggestive of increased π–π interactions between the pyrenyl residue of BPT and ^{Me}C:G or dPER:G containing DNA as compared to unmethylated DNA. The authors concluded that BPT assumes both quasi-external and intercalated geometries in its complexes with DNA. Furthermore, a greater contribution of intercalated structures was observed in the presence of ^{Me}C and dPER, as compared to an unmodified duplex. A similar trend is expected for DNA complexes with BPDE and other PAH diol-epoxides.

3.7. Separation of Molecules with Py Chromophore by Capillary Electrophoresis (CE) and High-Performance Liquid Chromatography (HPLC) with Online FLNS Detection

Nowadays separation methods like CE and HPLC are usually coupled with "information-rich" detection methods (or "hyphenated techniques"), such as, mass spectroscopy (MS), Raman, and nuclear magnetic resonance (NMR), for unambiguous online identification of separated analytes ([100] and references therein). Likewise, FLNS detection could also be interfaced with CE and CEC separation methods ([47] and references therein). The system setup is described in Figure 20. This system consists of a CE apparatus, a capillary cryostat (CC), a translation stage, laser excitation source, a monochromator, and either an intensified photodiode array or charge-coupled device for detection. Pulse generators are employed for time-resolved spectroscopy. The capillary inside CC can be cooled down by a flow of liquid nitrogen or liquid helium. This system allows rapid cooling to 4.2 K within 1 min and formation of glassy matrices for FLN detection. After freezing the analytes and buffer inside the capillary, arbitrary detection times will be used to completely characterize the analyte zones (black rectangles inside the capillary) by low-temperature NLN or FLNS.

Figure 21 [40] shows an example of applying this technique to separate different kinds of PAH molecules. In this analysis, a mixture of five PAHs—B[*a*]P, Py, 1-hydroxypyrene, benzo[*a*]anthracene (BA), and benzo[*e*]pyrene, is separated and identified by CE with online-coupled FLNS detection. An electropherogram is shown in the inset of Figure 21A, and it's generated by the three-dimensional plots of NLN fluorescence spectra versus capillary position. FLN spectrum for each analyte measured online is compared with the reference spectrum obtained offline for identification. This technique is also applied in the separation and identification of stable [43] and depurinating [42] PAH-DNA adducts. In addition, ref. [44] showed a successful application of the method to separate PAH-DNA adducts in human urine.

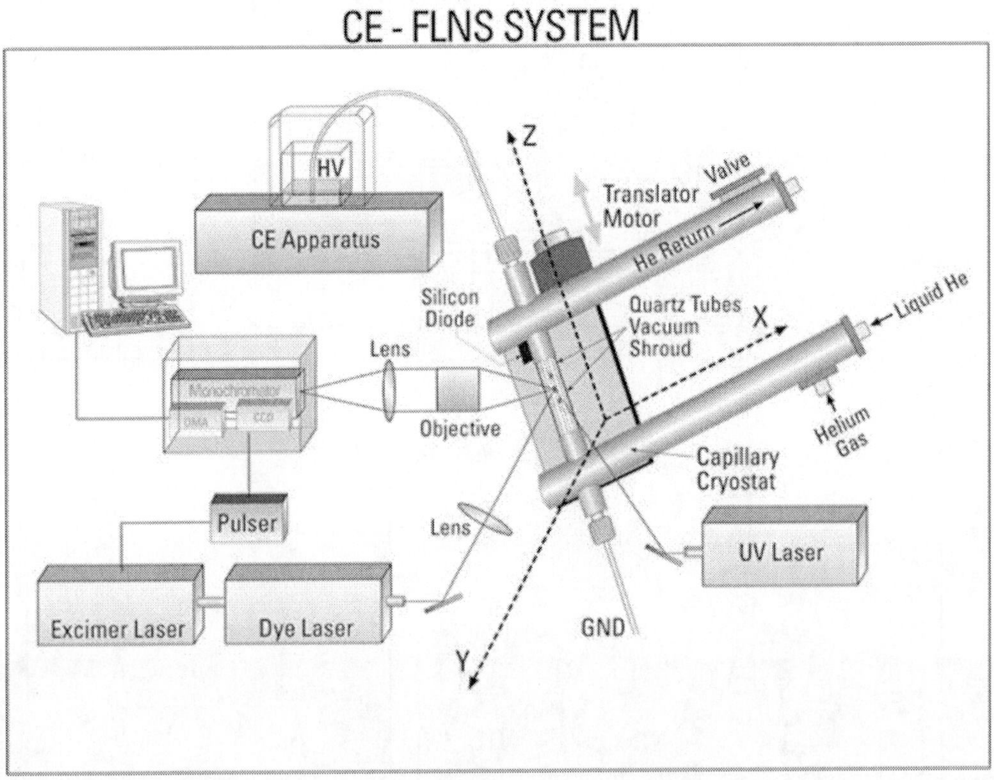

Figure 20. Schematic diagram of the CE–FLN system used for identification and structural characterization of molecular analytes. Reprinted with permission from [43].

In ref. [20], Roberts *et al.* reported that various analytes can be pre-separated by HPLC and detected on-line by FLN spectroscopy. The apparatus was similar to that shown in Figure 20, with the exception that the CE system was replaced by an HPLC instrument. For example, in this application, HPLC-FLNS successfully identified *trans*-7,8- and *trans*-9,10-B[*a*]P-diols, which co-eluted and couldn't be clearly resolved by HPLC alone. It was suggested that the HPLC-FLNS approach could be a useful tool in the analysis of complex mixtures where structurally similar analytes are commonly encountered. Situations such as these often arise in analyzing biological mixtures (e.g., depurinating PAH-DNA adducts and various metabolites formed *in vivo*), where absolute, unambiguous identification is of vital importance, and identification based solely on retention time may not be sufficient. With femtomole detection limits, both CE-FLNS and HPLC-FLNS methodologies are adequate for many real-world applications.

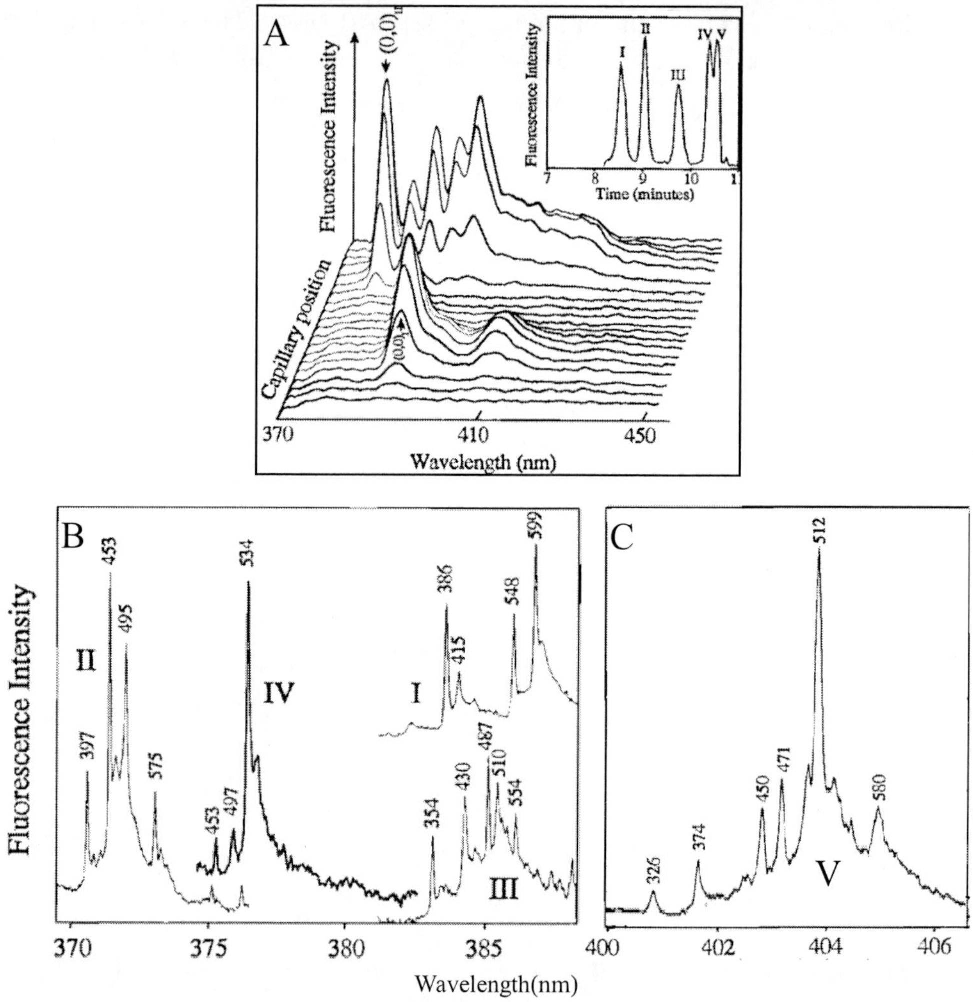

Figure 21. Frame A: Three-dimensional plot of NLN fluorescence spectra obtained by translating the capillary through the laser excitation and fluorescence observation region for 1-hydroxypyrene (I) and pyrene (II). Each trace represents a 1 s integrated fluorescence signal. Fluorescence (0,0)-origin bands are indicated. Inset: Fluorescence electropherogram for the PAH mixture, generated from the three-dimensional plots of NLN fluorescence versus capillary position. The labeled peaks are 1-hydroxypyrene (I), pyrene (II), B[*a*]A (III), B[*e*]P (IV), and B[*a*]P (V). Frames B and C: FLN spectra of the CE-separated PAHs. For analytes I to V, laser excitation wavelengths are 365.2, 369.0, 378.0, 378.0 and 395.7 nm, respectively. Reprinted with permission from [40]. Copyright 1996 American Chemical Society.

3.8. Biosensor Chips with FLNS Detection

New concepts for highly selective sensing and identification of analytes in complex mixtures are required in the area of advanced sensor development for the detection of toxic chemicals and various biological agents. For example, Duhachek *et al.* have developed an mAb-gold biosensor chip with low-T laser-induced fluorescence detection for analysis of

DNA-carcinogen adducts [101]. Their improved chip architecture [102] had a protein linker dithiobis(succinimidyl propionate) (DSP) attached to the gold surface, to which a recombinant protein A was attached to provide oriented immobilization of the mAbs. The schematic of major construction steps is illustrated in Figure 22A (see figure caption for details). The use of DSP led to faster protein-binding kinetics and higher protein surface density, while the incorporation of recombinant protein A for directed immobilization of mAbs increased the distance between mAb-bound analyte(s) and the gold surface, leading to improved detection limits [102]. This was because the increased distance minimized fluorescence quenching, resulting in about a ten-fold increase in the fluorescence signal in comparison to a previous chip design [101, 102]. The current limit of detection is in the low femtomole range. It was demonstrated that the improved chip architecture could be used for efficient biosensing of various DNA adducts. An example of an FLN spectrum of BPDE-10-N^2-dG adduct, measured directly on a biosensor chip at 4.2 K, is shown in Figure 22B. FLN spectra a and b were obtained with λ_{ex} of 367.0 and 363.0 nm, respectively. ZPLs are labeled with their excited-state vibrational frequencies (in cm^{-1}). It has been argued that biosensing of several structurally similar PAH and/or PAH-derived DNA adducts, bound to specific mAbs immobilized from a mixture at the same address on the chip, is feasible [101, 102].

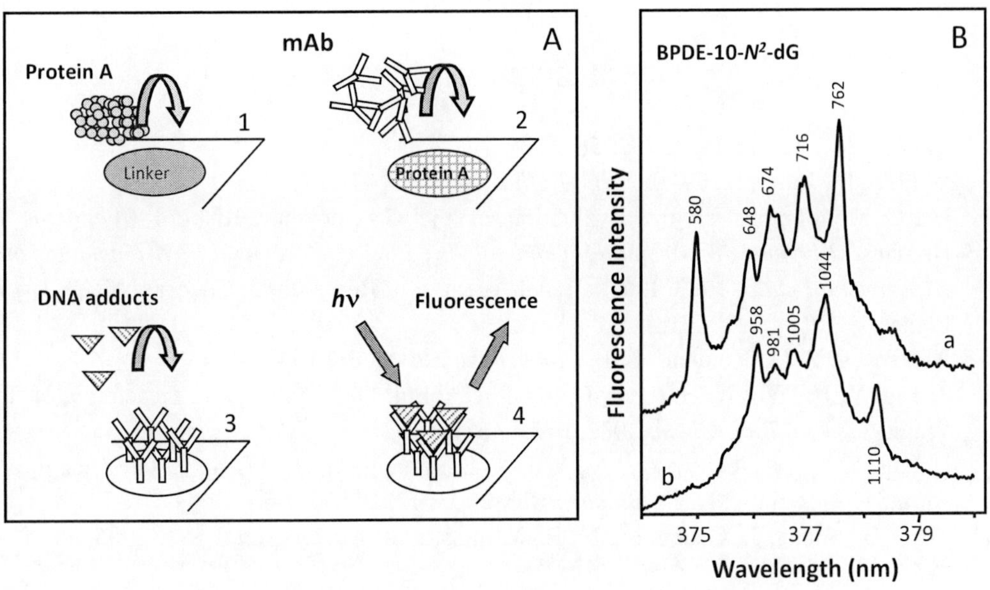

Figure 22. Frame A: Construction of the mAbs/protein A/DSP-gold biosensor chip: step 1, immobilization of protein A on the DSP linker; step 2, directed immobilization of mAbs on protein A; step 3, analytes of interest are captured by mAbs and the surface is covered with a thin layer of glycerol; step 4, spectroscopic interrogation by low-temperature, laser-based spectroscopy. Frame B: FLN spectra of BPDE-10-N^2-dG adducts on chip. Spectra a and b were obtained with λ_{ex} of 367.0 and 363.0 nm, respectively. ZPLs are labeled with their excited-state vibrational frequencies (in cm^{-1}). T = 4.2 K. Reprinted with permission from [101].

Conclusion

This chapter focuses on the basic aspects of low-temperature laser-induced fluorescence spectroscopy (under non-line-narrowing and line-narrowing conditions) and application of this approach to various biological systems containing Py and/or pyrenyl chromophore. A combination of fluorescent properties of these chromophores with FLN/NLN spectroscopies provides a powerful research tool in various areas. For example, one can characterize mAb-hapten interaction, identify the binding site in various PAH-derived DNA adducts, as well as study their conformations and interactions with a local environment. FLN spectroscopy can identify various stereoisomers, especially when separation techniques (CE and HPLC) are combined with on-line FLNS detection. Both online hybrids (i.e., CE-FLNS and HPLC-FLNS) can be successfully used to separate and analyze closely related analytes, even when they are formed at very low levels (e.g., detection limit of one adduct in 10^8 base pairs for 100 μg of macromolecular DNA [103], or femtomole of a moderately strong molecule such as B[*a*]P [11, 20, 103]). It is anticipated that fluorescence-based techniques (using highly fluorescent chromophores) will continue to be successful in structural identification and in studies related to the mechanisms of chemical carcinogenesis.

References

[1] Cook, J. W., *Nature* 1933, *131*, 56-57.

[2] Wild, S. R., Jones, K. C., *Environ. Pollut.* 1995, *88*, 91-108.

[3] IARC Monographs on the Evaluation of the Carcinogenic Risk of Chemicals to Humans: *Polynuclear Aromatic Compounds, Part I, Chemical,Environmental and Experimental Data*, 1983, International Agency for Research on Cancer (IARC), Lyon, France.

[4] Alekseyev, Y. O., Romano, L. J., *Biochemistry* 2000, *39*, 10431-10438.

[5] Zhang, Y. B., Wu, X. H., Guo, D. Y., Rechkoblit, O., Taylor, J. S., Geacintov, N. E., Wang, Z. G., *J. Biol. Chem.* 2002, *277*, 44582-44587.

[6] Huang, X. W., Kolbanovskiy, A., Wu, X. H., Zhang, Y. B., Wang, Z. G., Zhuang, P., Amin, S., Geacintov, N. E., *Biochemistry* 2003, *42*, 2456-2466.

[7] Zhao, B., Wang, J., Geacintov, N. E., Wang, Z. G., *Nucleic Acids Res.* 2006, *34*, 417-425.

[8] Ramakrishna, N. V. S., Gao, F., Padmavathi, N. S., Cavalieri, E. L., Rogan, E. G., Cerny, R. L., Gross, M. L., *Chem. Res. Toxicol.* 1992, *5*, 220-226.

[9] Rogan, E. G., Cavalieri, E. L., Tibbels, S. R., Cremonesi, P., Warner, C. D., Nagel, D. L., Tomer, K. B., Cerny, R. L., Gross, M. L., *J. Am. Chem. Soc.* 1988, *110*, 4023-4029.

[10] Jankowiak, R., Small, G. J., *Anal. Chem.* 1989, *61*, 1023 A -1032A.

[11] Jankowiak, R., Small, G. J., *Chem. Res. Toxicol.* 1991, *4*, 256-269.

[12] Personov, R. I., in: Agranovich, V. M., Hochstrasser, R. M. (Ed.), *Spectroscopy and Excitation Dynamic of Condensed Molecular Systems*, North-Holland, Amsterdam 1983, pp. 555-619.

[13] Hofstrat, J. W., Gooijer, C., Velthorst, N. H., in: Schulman, S. G. (Ed.), *Molecular Luminescence Spectroscopy: Methods and Applications*, John Wiley and Sons, New York 1988, pp. 383-459.

[14] Fidy, J., Vanderkooi, J. M., in: Douglas, R. H., Moan, J., Ronto, G. (Ed.), *Light in Biology and Medicine*, Plenum Press, New York 1991, pp. 367-374.

[15] Weber, M. J., *J. Lum.* 1987, *36*, R3-R4.

[16] Price, B. B., Wright, J. C., *Anal. Chem.* 1990, *62*, 1989-1994.

[17] Riesen, H., *Commet. Inorg. Chem.* 1993, *14*, 323-347.

[18] Jankowiak, R., in: Gooijer, C., Ariese, F., Hofstraat, J. W. (Ed.), *Shpol'skii Spectroscopy and Other Site-Selection Methods*, John Wiley and Sons, Inc., New York 2000, pp. 235-271.

[19] Ariese, F., Bader, A. N., Gooijer, C., *TRAC-Trend. Anal. Chem.* 2008, *27*, 127-138.

[20] Roberts, K. P., Jankowiak, R., Small, G. J., *Anal. Chem.* 2001, *73*, 951-956.

[21] Bowman, E. D., Rothman, N., Hackl, C., Santella, R. M., Weston, A., *Biomarkers* 1997, *2*, 321-327.

[22] Rogan, E. G., Ramakrishna, N. V. S., Higginbotham, S., Cavalieri, E. L., Jeong, H., Jankowiak, R., Small, G. J., *Chem. Res. Toxicol.* 1990, *3*, 441-444.

[23] Kok, S. S., Bakker, I., Gooijer, C., Brinkman, U. A. T., Velthorst, N. H., *Anal. Chim. Acta.* 1999, *389*, 77-83.

[24] Kok, S. J., *PhD Thesis*, Vrije Universitieit, de Boelelaan, Amsterdam, the Netherlands 1999.

[25] Kok, S. J., Posthumus, R., Bakker, I., Gooijer, C., Brinkman, U. A. T., Velthorst, N. H., *Anal. Chim. Acta.* 1995, *303*, 3-10.

[26] Cooper, R. S., Jankowiak, R., Hayes, J. M., Lu, P. Q., Small, G. J., *Anal. Chem.* 1988, *60*, 2692-2694.

[27] Hofstraat, J. W., Engelsma, M., Vandenesse, R. J., Gooijer, C., Velthorst, N. H., Brinkman, U. A. T., *Anal. Chim. Acta.* 1986, *186*, 247-259.

[28] Hofstraat, J. W., Gooijer, C., Velthorst, N. H., *Appl. Spectrosc.* 1988, *42*, 614-619.

[29] Vandenesse, R. J., Hoogland, G. J. M., Demoel, J. J. M., Gooijer, C., Brinkman, U. A. T., Velthorst, N. H., *J. Chromatogr.* 1991, *552*, 613-623.

[30] Vandenesse, R. J., Vinkenburg, I. H., Jonker, R. H. J., Hoornweg, G. P., Gooijer, C., Brinkman, U. A. T., Velthorst, N. H., *Appl. Spectrosc.* 1994, *48*, 788-795.

[31] Randerath, K., Reddy, M. V., Gupta, R. C., *P. Natl. Acad. Sci.-Biol.* 1981, *78*, 6126-6129.

[32] Meehan, T., Straub, K., *Nature* 1979, *277*, 410-412.

[33] Cheng, S. C., Hilton, B. D., Roman, J. M., Dipple, A., *Chem. Res. Toxicol.* 1989, *2*, 334-340.

[34] Geacintov, N. E., Cosman, M., Hingerty, B. E., Amin, S., Broyde, S., Patel, D. J., *Chem. Res. Toxicol.* 1997, *10*, 111-146.

[35] Marsch, G. A., *PhD Thesis*, Florida State University, Tallahassee, Florida, USA 1990.

[36] Marsch, G. A., Jankowiak, R., Farhat, J. H., Small, G. J., *Anal. Chem.* 1992, *64*, 3038-3044.

[37] Marsch, G. A., Jankowiak, R., Suh, M., Small, G. J., *Chem. Res. Toxicol.* 1994, *7*, 98-109.

[38] Yang, J. L., Maher, V. M., McCormick, J. J., *P. Natl. Acad. Sci.USA.* 1987, *84*, 3787-3791.

[39] Yang, J. L., Maher, V. M., McCormick, J. J., *Mol. Cell Biol.* 1987, *7*, 1267-1270.
[40] Jankowiak, R., Zamzow, D., Ding, W., Small, G. J., *Anal. Chem.* 1996, *68*, 2549-2553.
[41] Zamzow, D., Small, G. J., Jankowiak, R., *Mol. Cryst. Liq. Crys. A.* 1996, *291*, 155-162.
[42] Zamzow, D., Lin, C. H., Small, G. J., Jankowiak, R., *J. Chromatogr. A.* 1997, *781*, 73-80.
[43] Roberts, K. P., Lin, C. H., Jankowiak, R., Small, G. J., *J. Chromatogr. A.* 1999, *853*, 159-170.
[44] Roberts, K. P., Lin, C. H., Singhal, M., Casale, G. P., Small, G. J., Jankowiak, R., *Electrophoresis* 2000, *21*, 799-806.
[45] Ariese, F., Small, G. J., Jankowiak, R., *Carcinogenesis* 1996, *17*, 829-837.
[46] Landers, J. P., *Clin. Chem.* 1995, *41*, 495-509.
[47] Jankowiak, R., Roberts, K. P., Small, G. J., *Electrophoresis* 2000, *21*, 1251-1266.
[48] Jankowiak, R., Rogan, E. G., Cavalieri, E. L., *J. Phys. Chem. B.* 2004, *108*, 10266-10283.
[49] Telfer, A., Pascal, A. A., Bordes, L., Barber, J., Robert, B., *J. Phys. Chem. B.* 2010, *114*, 2255-2260.
[50] Rebane, K. K., Avarmaa, R. A., *Chem. Phys.* 1982, *68*, 191-200.
[51] Avarmaa, R., Jaaniso, R., Mauring, K., Renge, I., Tamkivi, R., *Mol. Phys.* 1986, *57*, 605-621.
[52] Pascal, A., Peterman, E., Gradinaru, C., van Amerongen, H., van Grondelle, R., Robert, B., *J. Phys. Chem. B.* 2000, *104*, 9317-9321.
[53] Renge, I., Mauring, K., Sarv, P., Avarmaa, R., *J. Phys. Chem.US.* 1986, *90*, 6611-6616.
[54] Rätsep, M., Linnanto, J., Freiberg, A., *J. Chem. Phys.* 2009, *130*, 194501-194501.
[55] Rätsep, M., Cai, Z.-L., Reimers, J. R., Freiberg, A., *J. Chem. Phys.* 2011, *134*, 024506-024506.
[56] Wendling, M., Pullerits, T., Przyjalgowski, M. A., Vulto, S. I. E., Aartsma, T. J., van Grondelle, R., van Amerongen, H., *J. Phys. Chem. B.* 2000, *104*, 5825-5831.
[57] Jankowiak, R., Reppert, M., Zazubovich, V., Pieper, J., Reinot, T., *Chem. Rev.* 2011, *111*, 4546-4598.
[58] Friedrich, J., Haarer, D., *Angew. Chem., Int. Ed. Engl.* 1984, *23*, 113-140.
[59] Jankowiak, R., Hayes, J. M., Small, G. J., *Chem. Rev.* 1993, *93*, 1471-1502.
[60] Moerner, W. E., *Topics in Current Physics. Persistent Spectral Hole Burning: Science and Applications.*, Springer-Verlag, New York 1988.
[61] Rebane, K. K., *Impurity Spectra of Solids*, Plenum, New York 1970.
[62] Völker, S., in: Fünfschilling, J. (Ed.), *Relaxation Processes in Molecular Excited States*, Kluwer Academic Publishers, Dordrecht 1989, p. 113.
[63] Stoneham, A. M., *Rev. Mod. Phys.* 1969, *41*, 82-108.
[64] Skinner, J. L., Moerner, W. E., *J. Phys. Chem.US.* 1996, *100*, 13251-13262.
[65] Osadko, I. S., in: Dusek, K. (Ed.), *Adv. Polym. Sci.*, Springer-Verlag, Berlin 1994, p. 123.
[66] Sapozhnikov, M. N., *Phys. Status. Solidi. B. Basic Research.*, 1976, *75*, 11-51.
[67] Dushinsky, F., *Acta Physiochem.* 1937, *7*, 551-557.
[68] Bogner, U., Schwarz, R., *Phys. Rev. B.* 1981, *24*, 2846-2849.
[69] Yaaniso, R. V., *Proc. Acad. Sci. Estonian SSR Phys. Math.* 1985, *34*, 277-283.
[70] Yaaniso, R. V., *J. Appl. Spectrosc.*1986, *44*, 365-370.
[71] Fünfschilling, J., Glatz, D., Zschokkegranacher, I., *J. Lum.* 1986, *36*, 85-92.

[72] Rätsep, M., Freiberg, A., *Chem. Phys. Lett.* 2003, *377*, 371-376.

[73] Timpmann, K., Rätsep, M., Hunter, C. N., Freiberg, A., *J. Phys. Chem. B.* 2004, *108*, 10581-10588.

[74] Freiberg, A., Rätsep, M., Timpmann, K., Trinkunas, G., *Chem. Phys.* 2009, *357*, 102-112.

[75] Rätsep, M., Pajusalu, M., Freiberg, A., *Chem. Phys. Lett.* 2009, *479*, 140-143.

[76] Rätsep, M., Pieper, J., Irrgang, K.-D., Freiberg, A., *J. Phys. Chem. B.* 2008, *112*, 110-118.

[77] Rätsep, M., Freiberg, A., *J. Lum.* 2007, *127*, 251-259.

[78] Shinohara, H., Yamakita, Y., Ohno, K., *J. Mol. Struct.* 1998, *442*, 221-234.

[79] Lin, C., Feng, X., Dang, N., Acharya, K., Jankowiak, R., unpublished results.

[80] Renge, I., Ratsep, M., Freiberg, A., *J. Lum.* 2011, *131*, 262-265.

[81] Reppert, M., Naibo, V., Jankowiak, R., *J. Chem. Phys.* 2010, *133*, 014506-014506.

[82] Nakajima, A., *B. Chem. Soc. JPN.* 1971, *44*, 3272-3277.

[83] Kalyanasundaram, K., Thomas, J. K., *J. Am. Chem. Soc.* 1977, *99*, 2039-2044.

[84] Kim, S. K., Brenner, H. C., Soh, B. J., Geacintov, N. E., *Photochem. Photobiol.* 1989, *50*, 327-337.

[85] Jankowiak, R., Lu, P. Q., Small, G. J., Geacintov, N. E., *Chem. Res. Toxicol.* 1990, *3*, 39-46.

[86] Grubor, N. M., Hayes, J., Small, G. J., Jankowiak, R., *Natl. Acad. Sci.USA.* 2005, *102*, 7453-7458.

[87] Lin, C., Chinnappan, R., Acharya, K., Pellequer, J.-L., Jankowiak, R., *Biophys. Chem.* 2011, *154*, 35-40.

[88] Chen, S. W. W., Pellequer, J. L., *Curr. Med. Chem.* 2004, *11*, 595-605.

[89] Pellequer, J. L., Zhao, B. T., Kao, H. I., Bell, C. W., Li, K., Li, Q. X., Karu, A. E., Roberts, V. A., *J. Mol. Biol.* 2000, *302*, 691-699.

[90] Chen, L., Devanesan, P. D., Higginbotham, S., Ariese, F., Jankowiak, R., Small, G. J., Rogan, E. G., Cavalieri, E. L., *Chem. Res. Toxicol.* 1996, *9*, 897-903.

[91] Koreeda, M., Moore, P. D., Wislocki, P. G., Levin, W., Conney, A. H., Yagi, H., Jerina, D. M., *Science* 1978, *199*, 778-780.

[92] Sims, P., Grover, P. L., Swaislan.A, Pal, K., Hewer, A., *Nature* 1974, *252*, 326-328.

[93] Cavalieri, E. L., Rogan, E. G., Devanesan, P. D., Cremonesi, P., Cerny, R. L., Gross, M. L., Bodell, W. J., *Biochemistry* 1990, *29*, 4820-4827.

[94] Grubor, N. M., Liu, Y., Han, X. X., Armstrong, D. W., Jankowiak, R., *J. Am. Chem. Soc.* 2006, *128*, 6409-6413.

[95] Miksa, B., Chinnappan, R., Dang, N. C., Reppert, M., Matter, B., Tretyakova, N., Grubor, N. M., Jankowiak, R., *Chem. Res. Toxicol.* 2007, *20*, 1192-1199.

[96] Jankowiak, R., Zamzow, D., Stack, D. E., Todorovic, R., Cavalieri, E. L., Small, G. J., *Chem. Res. Toxicol.* 1998, *11*, 1339-1345.

[97] Kim, S. K., Geacintov, N. E., Brenner, H. C., Harvey, R. G., *Carcinogenesis* 1989, *10*, 1333-1335.

[98] Lu, P. Q., Jeong, H., Jankowiak, R., Small, G. J., Kim, S. K., Cosman, M., Geacintov, N. E., *Chem. Res. Toxicol.* 1991, *4*, 58-69.

[99] Guza, R., Kotandeniya, D., Murphy, K., Dissanayake, T., Lin, C., Giambasu, G., Lad, R., Wojciechowski, F., Amin, S., Sturla, S., York, D., Jankowiak, R., Jones, R., Tretyakova, N., *Nucleic Acids Res.* 2011, *39*, 3988-4006.

[100] Kok, S. J., Velthorst, N. H., Gooijer, C., Brinkman, U. A. T., *Electrophoresis* 1998, *19*, 2753-2776.

[101] Duhachek, S. D., Kenseth, J. R., Casale, G. P., Small, G. J., Porter, M. D., Jankowiak, R., *Anal. Chem.* 2000, *72*, 3709-3716.

[102] Grubor, N. M., Shinar, R., Jankowiak, R., Porter, M. D., Small, G. J., *Biosens. Bioelectron.* 2004, *19*, 547-556.

[103] Jankowiak, R., Small, G. J., in: Neilson, A. H. (Ed.), *The Handbook of Environmental Chemistry: PAHs and Related Compounds*, Springer-Verlag, Berlin, Heidelberg, New York 1998, pp. 119-145.

In: Pyrene
Editors: Petr Ruzicka and Tomas Kral

ISBN: 978-1-62417-291-5

Chapter 5

PHOTO-ENERGY AND ELECTRON DONATING ABILITY OF PYRENE

Kazutaka Hirakawa
Department of Basic Engineering (Chemistry), Faculty of Engineering,
Shizuoka University, Japan

ABSTRACT

Photoinduced electron transfer and energy transfer can be possible in the photoexcited pyrene to porphyrin. In the molecular system, which consists of the pyrene and the porphyrin, the electron transfer from the photoexcited pyrene to the ground state porphyrin moiety competes with the energy transfer from the pyrene to the porphyrin. These photophysical processes could be controlled by the solvent polarity and the distance between the pyrene and the porphyrin in the pyrene-phosphorus(V) porphyrin system.

Using a similar molecular system, these photophysical processes can be selected by the pH of the solvent. In the case of the direct connecting molecular system of the pyrene and the porphyrin, ultra-fast electron transfer was observed. Furthermore, the electron transition process of pyrene may be used for the control of the singlet oxygen generation photosensitized by the porphyrin moiety in a water-soluble pyrenylporphyrin.

1. INTRODUCTION

Pyrene is an important organic dye for photochemical study [1]. Pyrene and its derivatives have absorption bands of approximately 300 ~ 350 nm. The photo-excitation energy of pyrene can be transferred to other dyes, such as porphyrin [2-4], which can absorb visible regions.

The redox potential for one-electron oxidation of pyrene is relatively small. Therefore, electron transfer (ELT) from pyrene to other molecules is possible. Photo-energy transfer (ENT) and ELT are important processes for photochemistry and photobiology. The competition process of ELT and ENT from photoexcited pyrene to the ground state of the acceptor molecule is possible. In addition, the ELT from the ground state of pyrene to photoexcited acceptor molecules is also possible. This character of pyrene is unique and useful for photochemical research.

Furthermore, the ELT process of pyrene may be used for the study of medical application [5], such as photodynamic therapy (PDT) [6-8]. The target-selectivity of the drug for this photo-therapy can be controlled by ELT of the pyrene. In this chapter, several topics regarding the photochemical and photo-medicinal application of pyrene are introduced.

2. Photochemical and Electrochemical Properties of Pyrene

Pyrene is soluble in most organic solvents, such as ethanol, methanol, acetonitrile, dichloromethane, chloroform, cyclohexane, toluene, and benzene. In solution, pyrene has a relatively strong absorption band of approximately 300 ~ 350 nm.

The first singlet excited (S_1) state energy of pyrene is approximately 3.3 eV (Figure 1). Its fluorescence quantum yield is relatively strong and the fluorescence lifetime is relatively long in various solvents. The ENT is possible in the long lifetime of the S_1 state of pyrene to other molecules.

Visible-light dyes, such as porphyrins, can accept the excitation energy from pyrene. The redox potential of one electron oxidation of pyrene is approximately 1.20 V vs saturated calomel electrode (SCE). Therefore, in general, pyrene acts as an electron donor.

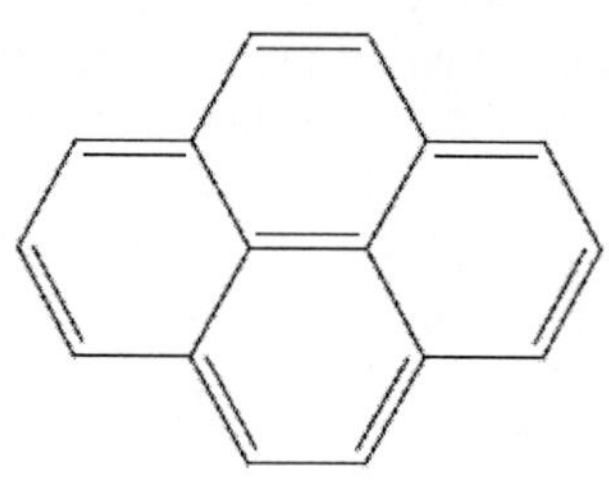

Excitation energy (singlet):	3.3 eV
Triplet energy:	1.8 eV
Redox potential (oxidation):	1.20 V vs SCE [9]
	1.25 V vs SCE
Fluorescence quantum yield:	0.72 (in ethanol) [10]
	0.49 (in dichloromethane)
	0.62 (in acetonitrile)
	0.52 (in toluene)
Fluorescence lifetime:	530 ns (in ethanol) [10]

Figure 1. Structure and photochemical properties of pyrene.

3. Competition of Electron Transfer and Energy Transfer from Pyrene to P(V)Porphyrin

3.1. Concept of the Pyrene-P(V)Porphyrin Triad System

In order to investigate the competition between ELT and ENT processes, a series of pyrene-porphyrin triads (T1 ~ T3) in which two pyrene moieties were connected at the central phosphorus atom of the porphyrin ring through various lengths of bridges, were used to investigate the interaction between the pyrene and the porphyrin moieties in the axial direction (Figure 2) [2]. The main reason for connecting two pyrenes to porphyrin is that ENT could be expanded to photo-energy harvesting [11, 12]. In the triads, the pyrene moiety acts as both an electron and an energy donor. As a result, ELT is considered to compete with ENT from photo-excited pyrene [13]. Since ELT and ENT are selected perfectly in a natural photosynthetic system, factors concerning the selectivity between ELT and ENT need to be discussed. With respect to this point, an attempt was made to explain the selectivity using classical theories.

3.2. Synthesis of Pyrene-P(V)Porphyrin Triads

These triads were synthesized by the following procedure, according to the previously reported method, from dichloroP(V)tetraphenylporphyrin chloride [14] and the corresponding 1-pyrene-alcohols, 1-hydroxypyrene, 1-pyrenemethanol, and 1-pyrenebutanol (Figure 3). The product was purified by column chromatography on silica gel using chloroform-methanol (10:1, v/v) as an eluent.

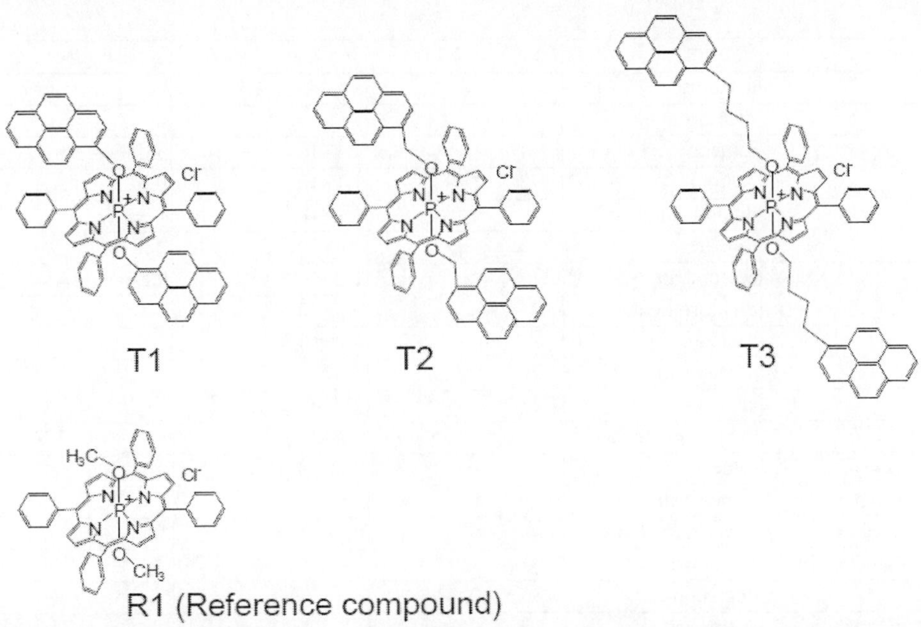

Figure 2. Structures of the twin-donor type pyrene-porphyrin triads.

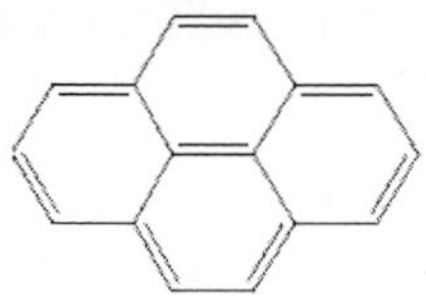

Pyrene

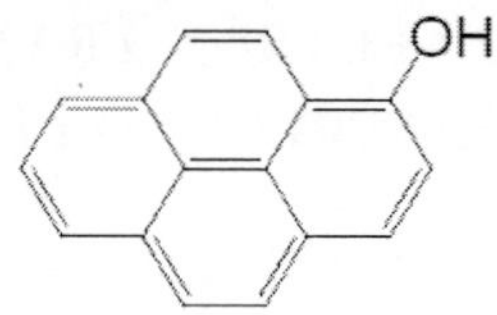

1-Hydroxypyrene

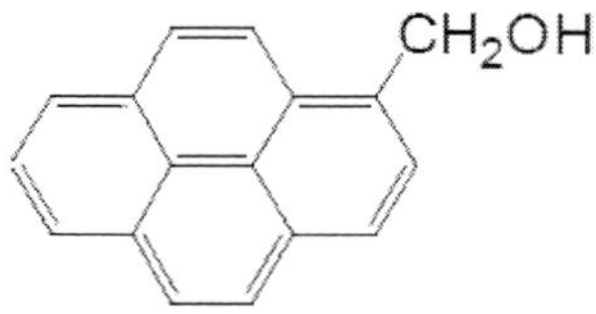

1-Pyrenemethanol

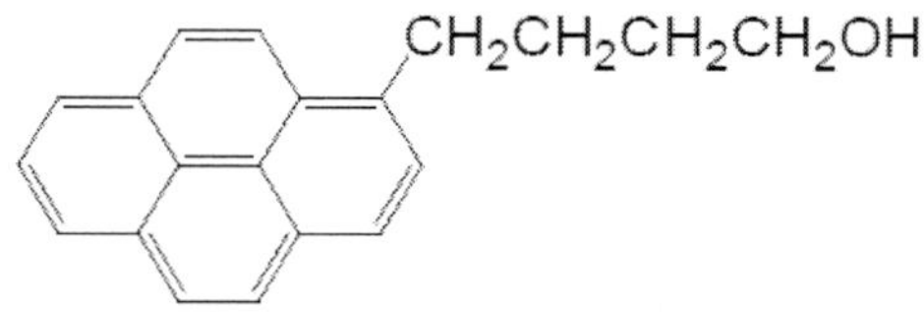

1-Pyrenebutanol

Figure 3. Structures of pyrene-alcohols used in this study.

3.3. Electrochemical and Photochemical Property of Pyrenyl Moieties of the Triads

The oxidation potentials of T1 and T2 were more positive than the corresponding reference pyrenes (Table 1).

Table 1. Redox potential and the fluorescence quantum yield and lifetime of the pyrene moiety of the triads and the reference pyrenes

Compound	Solvent	E_{ox} / V vs SCE	Φ_f	τ_f / ns
T1	acetonitrile	1.28	8×10^{-5}	nd
	dichloromethane		7×10^{-5}	nd
	toluene		1.1×10^{-4}	3.0
T2	acetonitrile	1.43	1.1×10^{-3}	0.6
	dichloromethane		1.1×10^{-3}	1.8
	toluene		4.4×10^{-3}	2.2
T3	acetonitrile	1.17	1.2×10^{-3}	0.5
	dichloromethane		9.3×10^{-4}	0.5
	toluene		5.0×10^{-3}	1.8
Pyrene	acetonitrile	1.20 *	0.62	
	acetonitrile	1.25		
	dichloromethane		0.49	
	toluene		0.52	
1-Hydroxypyrene	acetonitrile	1.09		
	dichloromethane		0.71	13
1-Pyrenemethanol	acetonitrile	1.31		
	benzonitrile	1.35		
1-Pyrenebutanol	acetonitrile	1.18		
	dichloromethane		0.51	60.6

*: [9].

Table 2. CT state energy and ELT property from the pyrene to the photoexcited porphyrin

Compound	Solvent	E_{Py*} / eV	E_{Por*} / eV	E_{CT} / eV	$k_{py \to por*}$ / s^{-1}
T1	acetonitrile	3.2	nd	1.7	too fast
	dichloromethane	3.2	nd	1.7	too fast
	toluene	3.2	2.0	1.8	4.4×10^{10}
T2	acetonitrile	3.3	2.0	1.8	1.1×10^{10}
	dichloromethane	3.3	2.0	1.8	1.8×10^{10}
	toluene	3.3	2.0	2.0	3.4×10^{8}
T3	acetonitrile	3.3	2.0	1.7	3.8×10^{9}
	dichloromethane	3.3	2.0	1.7	2.4×10^{9}
	toluene	3.3	2.0	1.9	1.9×10^{8}

E_{Py*}: S_1 energy of the pyrenyl moiety. E_{Por*}: S_1 energy of the porphyrin moiety.

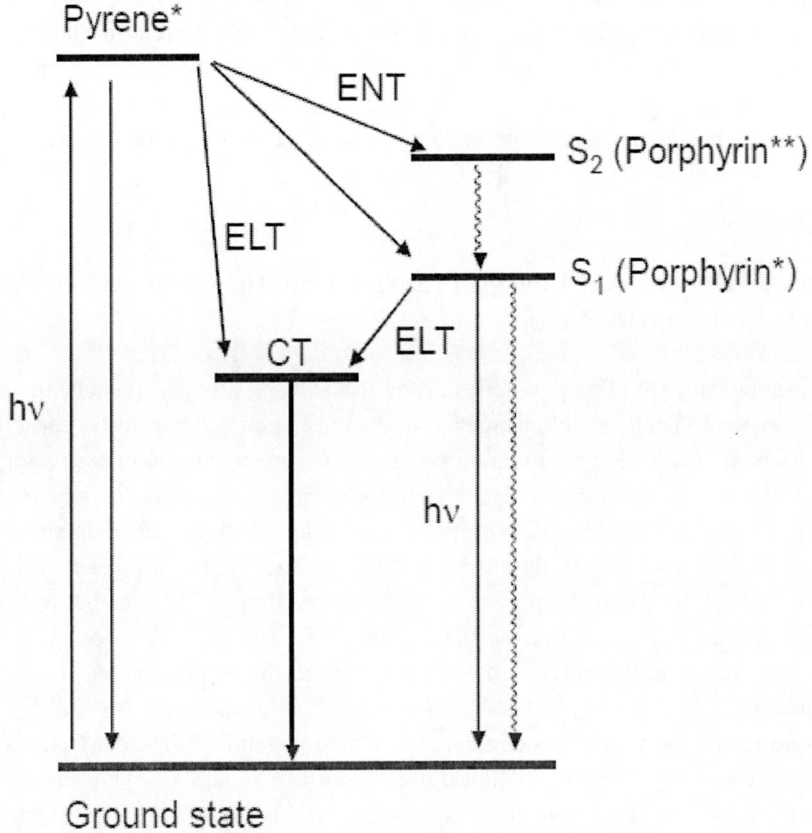

Figure 4. Energy levels of the photoexcited states of the pyrene and porphyrin and CT state.

These positive shifts of oxidation potentials indicate an electron donating character of the pyrene moieties in triads. In T3, a very small difference was observed, indicating a weak interaction in the ground state for the long distance between pyrene and porphyrin. The energy levels of the charge transfer (CT) states (E_{CT}) of triads in the solvents were estimated from the difference in the reduction potential of porphyrin and the oxidation potential of pyrene with ion-pair formation free enthalpy [15]. The estimated E_{CT} values are listed in

Table 2 with the energy of the S_1 states of pyrene (E_{py*}) and porphyrin (E_{por*}). The E_{CT} is lower than E_{por*} in acetonitrile and dichloromethane solutions. However, E_{CT} is very close to E_{por*} in toluene. These differences are caused by the stabilized energy of the CT states depending on solvent polarity. Schematic representations of the energy levels are given in Figure 4. These relationships of energy levels suggest the possibility of controlling ELT and ENT processes by solvent polarity. On the other hand, since the energy level of the porphyrin HOMO is lower than that of pyrene, the ELT from the ground state pyrene to the excited porphyrin is expected to occur when the porphyrin ring is photo-excited.

3.4. Electron Transfer from the Pyrenyl Moiety to the Photoexcited Porphyrin Ring

The fluorescence quantum yields of the pyrenyl moiety of the triads were smaller than that of R1 in any solvent, and became remarkably smaller as the pyrene-porphyrin distances (d) were reduced. The fluorescence quenching was explained by ELT from the pyrene to the photoexcited porphyrin. These ELT rate constants ($k_{py\rightarrow por*}$) were listed in Table 2. The $k_{py\rightarrow por*}$ of T1 is considerably large compared to those of T2 and T3. The $k_{py\rightarrow por*}$ of the triads decreased with increasing d and depended on the solvent.

3.5. Electron Transfer and Energy Transfer from the Photoexcited Pyrenyl Moiety to the Porphyrin Ring

The emission from the pyrene moieties of the triads were strongly quenched compared to the reference pyrene (Table 1). Fluorescence quenching is considered to be caused by ELT and ENT, or both, from excited pyrene. When ENT occurs, emission from porphyrin is observed by pyrene excitation, and a pyrene absorption structure is observed in the fluorescence excitation spectra of porphyrin emission. If these phenomena cannot be observed, only ELT quenching should be taken into account. Porphyrin emission was observed for T2 and T3 when the pyrene moiety was excited at 337 nm, but not for T1. In toluene, however, the emission from porphyrin rings of T2 and T3 was observed distinctly. The intensities were significantly higher than those of the emissions in acetonitrile and dichloromethane.

Furthermore, the excitation spectra have a structure similar to the pyrene absorption structure in the wavelength region from 300 to 360 nm for T2 and T3. These results suggest that ENT is the main process in toluene. The quantum yields of ENT from the excited pyrene to the porphyrin moieties were 0.67 and 0.83 for T2 and T3, respectively. Except for ENT, the relaxation is mainly ELT from the excited pyrene to the porphyrin. The ELT rate constant from pyrene ($k_{py*\rightarrow por}$) was estimated from the fluorescence quantum yields for the pyrene moiety and the reference pyrene and the fluorescence lifetime of the reference pyrene (Table 3). This value is considered to be dependent on the d, the free energy change, and the reorganization energy. Although $k_{py*\rightarrow por}$ increases with decreasing d, no linearity between the logarithm of $k_{py*\rightarrow por}$ and the d was observed. Apparent attenuation coefficients for T2 and T3 were 0.10 and 0.12 Å, respectively. These small values result from the partial overlap between the pyrene and the porphyrin moieties.

Table 3. ELT and ENT properties from the pyrene to the porphyrin

Compound	Solvent	$k_{py^* \to por}$ / s^{-1}	k_{ENT} / s^{-1}	Φ_{ENT}
T1	acetonitrile	2.9×10^{10}	nd	nd
	dichloromethane	2.3×10^{10}	nd	nd
	toluene	1.3×10^{10}	nd	nd
T2	acetonitrile	2.1×10^{9}	trace	trace
	dichloromethane	1.5×10^{9}	trace	trace
	toluene	1.1×10^{8}	2.3×10^{8}	0.67
T3	acetonitrile	1.9×10^{9}	trace	trace
	dichloromethane	1.8×10^{9}	trace	trace
	toluene	4.6×10^{7}	2.4×10^{8}	0.83

3.6. Competition Between Electron Transfer and Energy Transfer

The ELT rate can be controlled by solvents and the *d*. The solvent dependencies of the ENT rate constants (k_{ENT}) for excited pyrene are listed in Table 3. In the case of T1 in all solvents, energetically allowed ENT was inhibited by the strong interaction between the porphyrin and the pyrene, which enhanced the ELT. In T2 and T3 with relatively large *d*, ENT and ELT coexisted and competed as relaxation processes in toluene, whereas ELT was the main process in acetonitrile and dichloromethane. Since the energy levels of the CT state become higher in a less polar solvent, the energy levels of CT states in acetonitrile and dichloromethane (a relatively polar solvent) are lower than those in toluene (a relatively less polar solvent). In the case of photosynthetic reaction centers, it is known that surrounding a special pair are relatively less polar groups [16-18]. The less polar matrix is considered to be useful for the coexistence of ENT to the special pair and ELT from the special pair.

On the other hand, the energy levels cannot explain their distance dependencies, where the efficiency of ENT increases with increasing *d*. When ENT occurs through dipole-dipole interaction (Förster mechanism), the critical distance (R_0) is expressed by the following equation:

$$R_0^6 = \frac{9000 \ln 10 \kappa^2 \Phi^0_{Py}}{128\pi^5 N_A n^4} \int F(\nu)\varepsilon(\nu)\nu^{-4}\, d\nu \qquad (1)$$

where κ is geometrical factor, N_A is the Avogadro constant, *n* is the refractive index, $F(\nu)$ is the normalized fluorescence spectrum of the pyrene moiety, Φ^0_{Py} is the fluorescence quantum yield of the pyrene without quenching, $\varepsilon(\nu)$ is the absorption spectrum of the porphyrin ring, and ν is the wavenumber [19]. When the geometry was assumed to be random orientation, R_0 was estimated at about 60 Å. When assuming a Förster-type ENT, by which ENT was attenuated by d^{-6}, ENT became a major process in the region of large *d* because ELT was attenuated to an exponent of *d*.

In summary, emissions from the photoexcited states of pyrene and porphyrin were quenched through ELT and ENT, or both. In the relaxation from the photoexcited pyrene, the predominant process changed according to solvents and distances between the pyrene and the porphyrin moieties.

4. Control of the Electron and Energy Transfer from the Photoexcited Pyrene by pH

4.1. Concept and Molecular Design

The P(V) porphyrin is a strong electron acceptor by the central cationic phosphorus atom [20, 21]. The strength as an electron acceptor depends on the electron affinity of an axial ligand connected at the central phosphorus atom of the P(V)porphyrin. Therefore, a control of the condition of the axial ligand leads to the control of the electron affinity of the P(V) porphyrin and the control of ENT and ELT. For this purpose, a changeable axial ligand is considered to be a key component. In the case of the dihydroxy-P(V) porphyrin, the electronic condition of the π-electron system and the electron affinity is changed by acid dissociation of the axial hydroxyl groups [22]. This characteristic can be used for the external control of the electron affinity by pH of the surroundings. For this purpose, hydroxy(1-pyrenebutoxy)-P(V) tetrakis(*p*-butoxyphenyl)porphyrin chloride (Py-POH, Figure 5) was synthesized to investigate the external control of ENT and ELT from the pyrene moiety to the porphyrin ring by the change in conditions of the axial hydroxyl group [23].

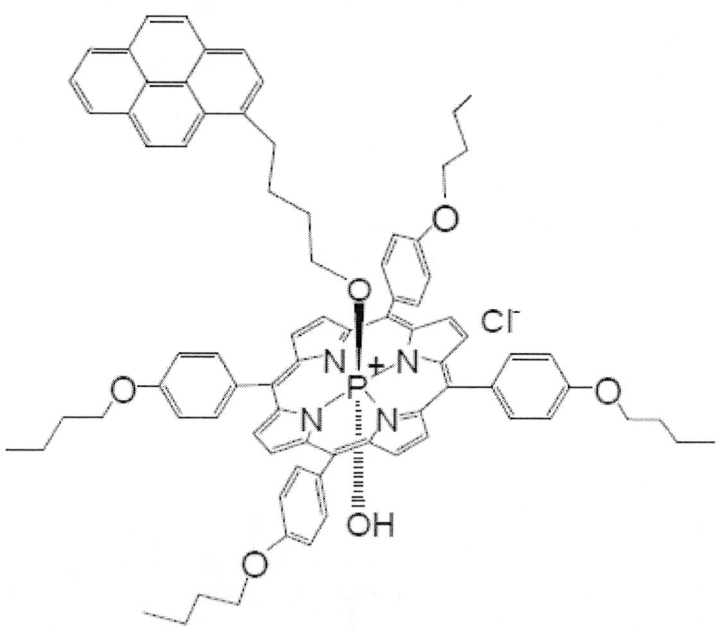

Figure 5. Structure of Py-POH.

4.2. Synthesis of Pyrene-Connecting P(V)Porphyrin

Py-POH was synthesized by the following procedure, according to the previously reported method from dichloroP(V) tetrakis(*n-p*-butoxyphenyl)porphyrin chloride [14], 1-pyrenebutanol, and silver nitrate [23]. The product was purified by column chromatography on silica gel, with chloroform-methanol (10:1, v/v) as an eluent.

4.3. Photophysical and Electrochemical Properties of Py-POH

Absorption spectrum of Py-POH in dichloromethane is shown in Figure 6. This spectrum represents almost superposition of the pyrene moiety (around 250 ~ 350 nm) and the Soret (400 ~ 470 nm) and the Q-bands (550 ~ 650 nm) of the porphyrin ring, suggesting that the photoexcited states of the pyrene and the porphyrin are localized on the corresponding moieties. Fluorescence spectra of Py-POH were assigned to those of the pyrene (350 ~ 450 nm) and the porphyrin (550 ~ 750 nm) moieties. Fluorescence from the pyrene moiety is remarkably quenched compared to reference pyrene, and the fluorescence from the porphyrin ring was clearly observed. The fluorescence excitation spectrum of this porphyrin fluorescence was fitted with the absorption spectrum of the pyrene moiety, indicating the ENT from the photoexcited pyrene to the porphyrin ring.

The reduction potential of the porphyrin ring and the oxidation potential of the pyrene moiety are -0.68 V and +1.20 V vs. SCE, respectively. The values of E_{CT} were estimated from these redox potentials using dielectric continuum theory [15]. The calculated values (eV) are 1.82 (acetonitrile), 1.97 (dichloromethane), and 1.81 (acetonitrile/water). The ELT from the photoexcited pyrene to the porphyrin is energetically possible, as is the ENT.

4.4. Control of the Electron and Energy Transfer from Pyrene to P(V)Porphyrin by pH

The fluorescence intensity of the pyrene moiety of Py-POH was markedly quenched by the porphyrin ring in acetonitrile/water (1:1, v/v). The ENT from the pyrene to the porphyrin was confirmed by the fluorescence excitation spectrum measurement. The ELT is also possible from energetic-point of view.

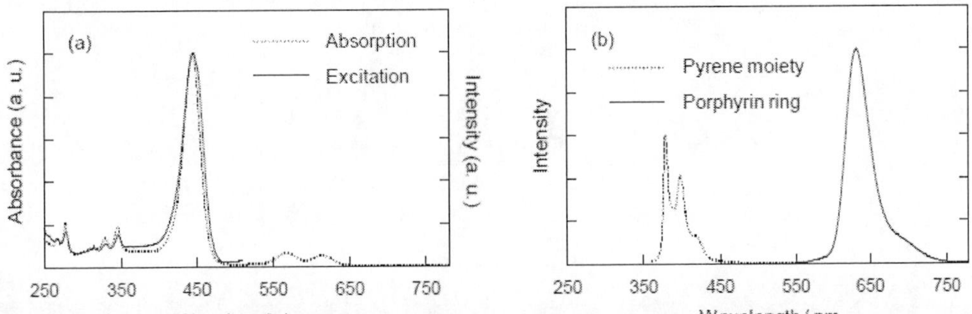

Figure 6. Absorption (a) and fluorescence (b) spectra of Py-POH.

It has been reported that the competition between the ENT and the ELT in the photoexcited pyrene-P(V) porphyrin system was possible [2]. The calculated critical distance of ENT from the photoexcited pyrene moiety to the P(V) porphyrin through dipole-dipole interaction (Förster mechanism) [19] is 42 Å. Because the center-to-center distance between the pyrene moiety and the porphyrin ring, which was estimated from the molecular mechanics calculation, is 8.5 Å, it makes sense that a high yield of ENT occurs in this molecular system. On the other hand, it was reported that the ELT rate constant from photoexcited pyrene to the tetraphenylporphyrinP(V) complex separated by butoxy chain is 1.8×10^{9} s^{-1} in dichloromethane, and that the ENT competed with the ELT [2].

The competition between the ENT and the ELT should be a reasonable phenomenon in the molecular system of Py-POH. These rate constants can be roughly estimated from the present experimental results. The roughly estimated k_{ENT} is around 5×10^{8} s^{-1}. The Φ_{ENT} should increase through the decrease of k_{ELT}, leading to the switching of these photophysical processes.

The P(V) porphyrin has a strong electron affinity due to the positive charge from a central phosphorus cation [24]. The pK_a of the axial hydroxyl group of Py-POH was estimated to be 9.5, and can be dissociated in alkali solution.

Fluorescence quantum yields of the porphyrin moiety increased depending on pH of the solvent, as shown in Figure 7.

These results can be reasonably explained by he electron affinity of the P(V) porphyrin having decreased through the neutralization of the positive charge of P(V) by the negative charge produced from the acid dissociation of the axial hydroxyl group, resulting in the suppression of the intramolecular ELT. This suppression of the ELT makes the ENT the predominant process.

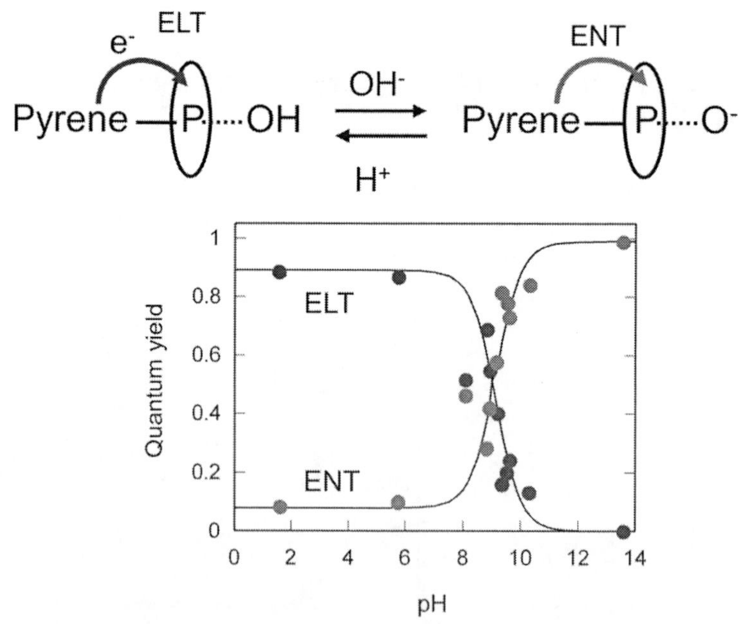

Figure 7. Scheme of the control of the ELT and ENT by the hydroxyl group of Py-POH.

5. Electron Transfer from Pyrene to the S_2 Excited State of Porphyrin

5.1. Design of Rapid Electron Transfer System Using Pyrene and Porphyrin

An enormous number of porphyrin molecular systems have been investigated on the ELT from the S_1 state, which is a notation of electronic states populated by excitation in the Q-band (500 ~ 650 nm) [25-28]. On the other hand, the porphyrin derivatives have a characteristic intense absorption, namely the Soret-band (400 ~ 450 nm) owing to a transition to upper excited states with a large energy gap.

The upper excited states have been denoted as the second excited singlet state (S_2), and some of the metalloporphyrins and diacid porphyrins emit an S_2 fluorescence, whose quantum yields [29-39] and lifetimes [33, 40-49] were in the region of 10^{-3} ~ 10^{-5} and 0.8 ~ 4.5 ps, respectively. Although the lifetimes of S_2 are remarkably shorter than those of S_1, intermolecular [40, 50] and intramolecular [43-49] ELT from S_2 within the S_2 lifetimes have been reported. When the ELT from S_2 occurs against internal conversion (IC) from S_2 to S_1, the ELT must be an ultra-fast process, which is an attractive phenomenon in both basic and applied photochemistry. The S_2 fluorescence is a useful probe to investigate the ultra-fast ELT from the upper excited state.

To investigate the rapid ELT from pyrene to porphyrin S_2, the following molecules, zinc *meso*-(1-pyrenyl)triphenylporphyrin derivatives (Figure 8) were synthesized [51].

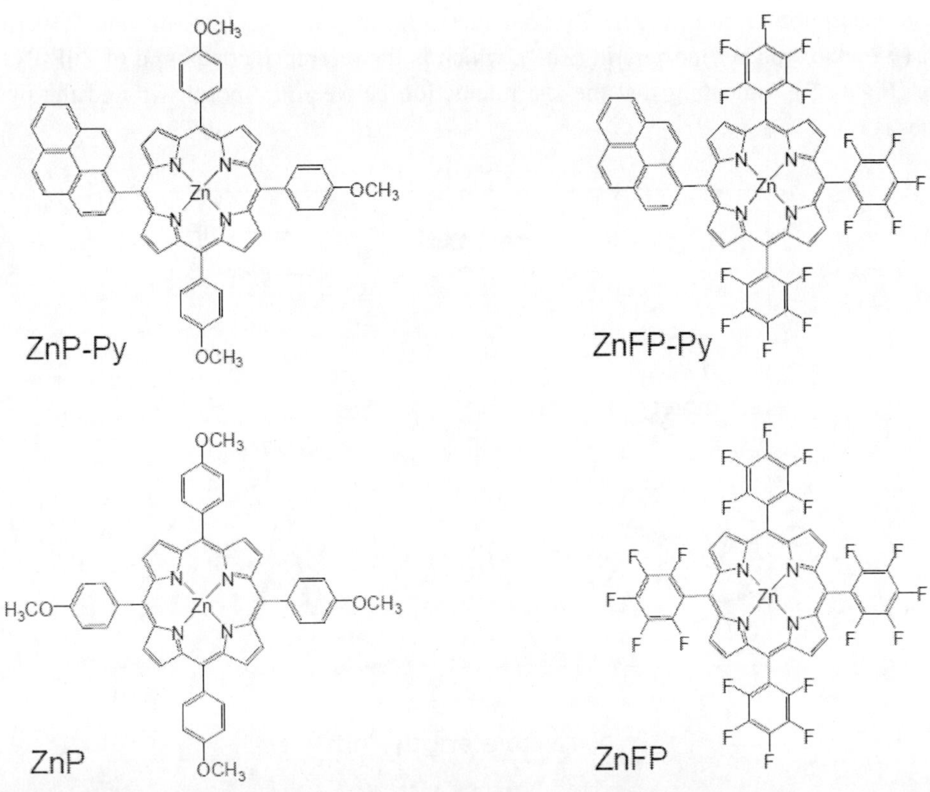

Figure 8. Structures of *meso*-pyrenylporphyrins and their reference porphyrins.

In the molecules, (i) direct linkage between porphyrin and pyrene enhances intramolecular ELT, (ii) the pyrene unit cannot become an energy acceptor of porphyrin S_2, and (iii) the energy level of the CT state is below S_2 and above S_1; therefore, ELT cannot occur from S_1. These molecules may be used as a model of photooxidative ELT to the porphyrin S_2. If a rapid ELT from S_2 occurs in these molecules, the S_2 fluorescence quenching becomes an effective probe of the ELT process under various conditions. Thus, the S_2 fluorescence quantum yields and its solvent dependence were investigated.

5.2. Synthesis of *Meso*-Pyrenylporphyrins

Zinc 5-(1'-pyrenyl)-10,15,20-tris(*p*-methoxyphenyl)porphyrin (ZnP-Py, Figure 8) and zinc 5-(1'-pyrenyl)-10,15,20-tris(pentafluorophenyl)-porphyrin (ZnFP-Py) were synthesized by the following procedures. At first, free-base *meso*-(1'-pyrenyl)triphenylporphyrin derivatives were synthesized from 1-pyrenecarboxaldehyde, *p*-methoxybenzaldehyde (or pentafluorobenzaldehyde), and pyrrole using a modified version of the procedure of Adler and coworkers [52]. Zinc ion was incorporated into the porphyrins using zinc acetate, according to the literature [53].

5.3. Absorption and Fluorescence Property of *Meso*-Pyrenylporphyrins

The absorption feature of ZnP-Py coincides with the superposition of zinc 5,10,15,20-tetrakis(*p*-methoxyphenyl)porphyrin (ZnP), which is the reference compound of ZnP-Py, and pyrene (Figure 9), indicating that the π-π interaction between the porphyrin and the pyrene moieties is weak.

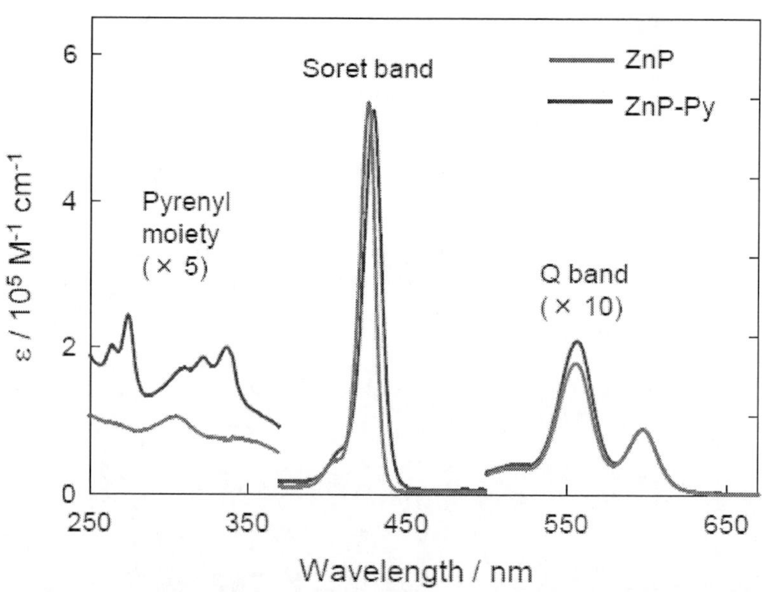

Figure 9. Absorption spectrum of *meso*-pyrenylporphyrin and the reference porphyrin. *e* indicates the molar absorption coefficient.

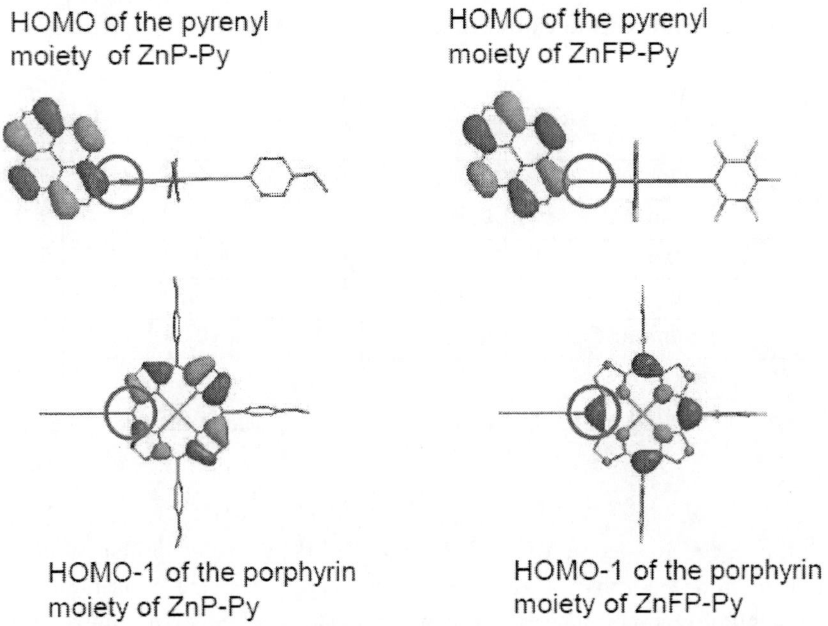

Figure 10. Calculated MO of *meso*-pyrenylporphyrins. The red circle indicates the 5-C position.

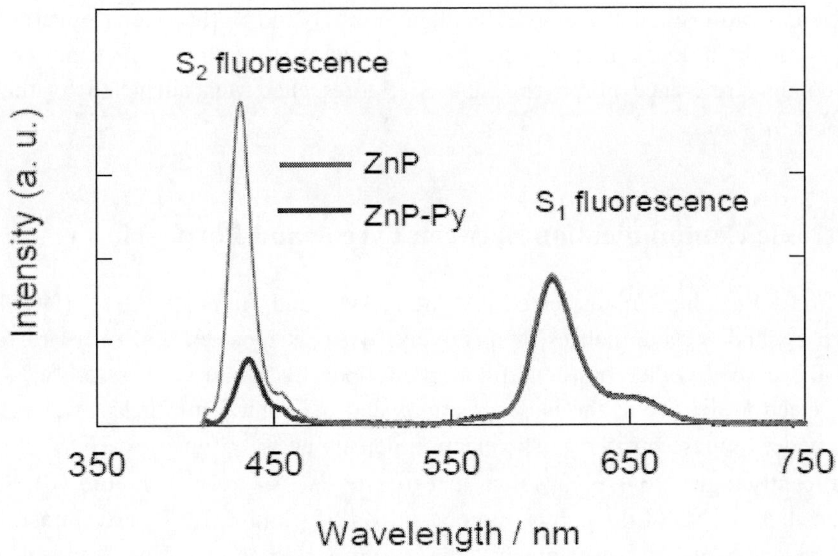

Figure 11. Fluorescence spectra of *meso*-pyrenylporphyrin and the reference porphyrin.

The weak π-π interaction is due to steric rotational hindrance of the pyrene moiety around the *meso*-position of the porphyrin, which keeps the two π-electronic systems nearly orthogonal to each other. The absorption feature of ZnFP-Py is similar to that of ZnP-Py. An *ab initio* MO calculation also indicated that the planes of the pyrene moieties are perpendicular to those of the porphyrins (Figure 10). This calculation showed that HOMO and LUMO of these porphyrins and pyrenes were localized on their corresponding moieties, supporting the results of absorption measurements.

Table 4. Fluorescence quantum yield of *meso*-pyrenylporphyrins

Compound	Solvent	f	$\Phi_{S2} \times 10^{-4}$	Φ/Φ_0 (S_2)	Φ/Φ_0 (S_1)
ZnP-Py	toluene	0.253	1.40	0.50	1.1
	dichloromethane	0.641	0.98	0.24	1.0
	acetonitrile	0.785	0.86	0.10	1.0
ZnFP-Py	hexane	0.187	0.44	0.42	1.0
	toluene	0.253	0.50	0.39	1.0
	dichloromethane	0.641	0.05	0.13	1.2
	acetonitrile	0.785	0.20	0.15	1.2

$f = \frac{2(\varepsilon - 1)}{2\varepsilon + 1} - \frac{n^2 - 1}{2n^2 + 1}$, where ε and n are dielectric constant and refractive index of solvents, respectively.

The S_1 fluorescence spectra of ZnP-Py and ZnFP-Py were similar to those of the corresponding reference porphyrins (Figure 11), indicating that the contribution of the CT state in S_1 is negligibly small. The fluorescence intensities of S_1 of the porphyrin moieties were almost the same as those of the corresponding reference porphyrins in any solvents when the porphyrins were photo-excited at the Q-bands. Consequently, it is confirmed that the S_1 of the porphyrin derivatives is not quenched by the pyrene moiety. When ZnP-Py and ZnFP-Py were photo-excited at the Soret band, both the S_2 and S_1 fluorescence were observed (Figure 11). Although the S_1 fluorescence quantum yields were almost the same as those of the corresponding reference porphyrins, the S_2 fluorescence was quenched by the pyrene moiety (Table 4).

5.4. Electronic Communication Between Pyrene and Porphyrin

Ab initio MO calculations indicated that the HOMO and the next HOMO (HOMO-1) of ZnP-Py are a_{2u} and a_{1u} symmetries, respectively, whereas those of ZnFP-Py are reversed, similar to the case of other tris(pentafluorophenyl)porphyrin derivatives [54-56]. The a_{2u} orbital has electron density at the *meso* positions and the central nitrogens, whereas the a_{1u} orbital has nodes at these positions. The electron density at 5-C of the next HOMO of ZnFP-Py is significantly higher (6.4×10^{-2}) than that of ZnP-Py (6.2×10^{-6}) (Figure 10). Since the electron densities at 1'-C of the pyrene moieties of ZnP-Py and ZnFP-Py are almost the same (8.9×10^{-2} and 8.5×10^{-2}, respectively), the electronic communication between the next HOMO of porphyrin and the HOMO of pyrene in ZnFP-Py becomes larger than that of ZnP-Py.

5.5. Electron Transfer from Pyrene to Porphyrin S_2

The S_2 fluorescence of the porphyrin is quenched by the pyrene moiety, whereas the S_1 fluorescence is not. Since the energy levels of the excited state of the pyrene moiety and the CT state of these pyrenylporphyrins are higher than that of S_1 of the porphyrin, quenching through ENT and ELT does not occur from the S_1. In the S_2, ELT from the pyrene moiety to the porphyrin S_2 is energetically possible, because the energy level of the CT state of a porphyrin anion and a pyrene cation pair is located between the S_2 and the S_1 energy levels of the porphyrins. The transient absorption spectrum measurement showed the absorption peak to be approximately 460 nm. This absorption could be assigned to the radical cation of pyrene [57], and did not coincide with the reported absorption peak of the radical anion of pyrene [58], strongly suggesting ELT from the pyrene to porphyrin S_2. If the S_2 quenching is due to the ELT, dependence of the fluorescence quenching on the solvent polarity is expected. The fluorescence quantum yields ratio (Φ/Φ_0), which results from the division of the fluorescence quantum yields of ZnP-Py and ZnFP-Py by those of the corresponding reference porphyrins, was listed against the solvent polarity parameter (*f*) (Table 4). The Φ/Φ_0 of S_2 decreased with increasing the solvent polarity, whereas the Φ/Φ_0 of S_1 was not affected. If the S_2 quenching process is due to enhancement of direct IC to S_1, the solvent dependence of S_2 fluorescence cannot be explained. For the purpose of careful investigation, the following deactivation processes from the S_2 of ZnP-Py and ZnFP-Py were assumed: (1) Direct IC from S_2 to S_1, and (2) ET between pyrene and the porphyrin S_2, which is followed by charge recombination (CR) to the porphyrin S_1 or S_0 (Figure 12).

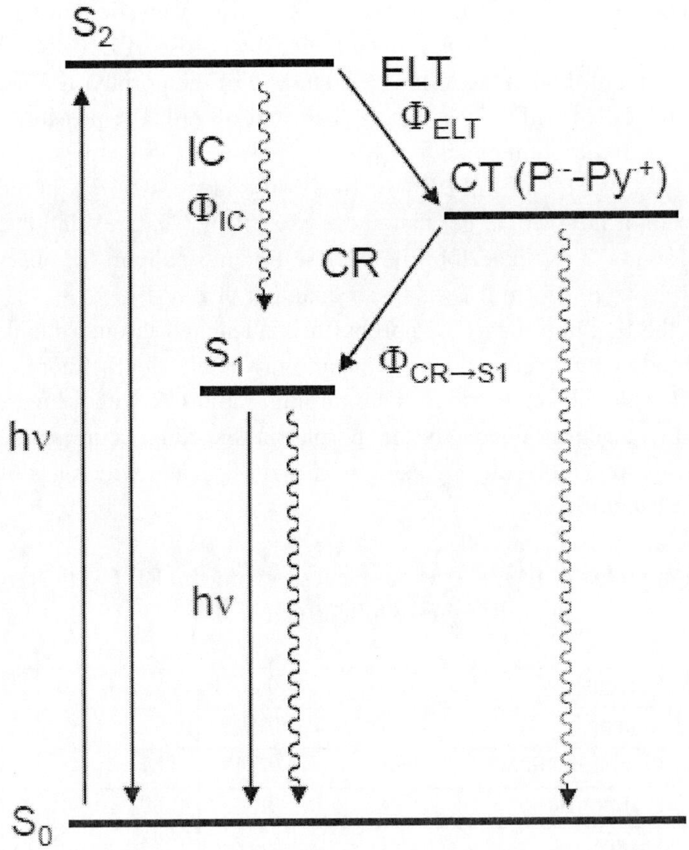

Figure 12. Relaxation process of the S_2 state of the porphyrin moiety of *meso*-pyrenylporphyrin.

Since the S_1 fluorescence quantum yields of reference porphyrins are independent of the excitation wavelengths within experimental error similar to that of previous reports [32, 33, 36], the main pathway of the deactivation processes of S_2 of these reference molecules are IC and S_2 emission.

Consequently, the quantum yields of the ELT (Φ_{ELT}), the IC (Φ_{IC}), and the CR to the porphyrin S_1 ($\Phi_{CR\rightarrow S1}$) can be estimated from the fluorescence intensity (Table 5). The Φ_{ELT} become large in polar solvent. For further investigation, the relative ELT rate constant (k_{ELT}) was roughly estimated from the following equation using the S_2 fluorescence quantum yields of the pyrenylporphyrin (Φ_{S2}) and the reference porphyrin ($\Phi^0{}_{S2}$):

$$k_{ELT} = \left(\frac{1}{\Phi_{S2}} - \frac{1}{\Phi^0{}_{S2}} \right) k_f \tag{2}$$

where k_f is the rate constant of S_2 fluorescence (3.5 ps). The k_{ELT}s mostly increase with increasing the solvent polarity almost coinciding with the dependency of the Φ/Φ_0 on the solvent polarity. The k_{ELT} value of ZnFP-Py in dichloromethane is larger than that in relatively high polar acetonitrile. This complicated result might be explained by the effect of charge transfer interaction between the porphyrin S_2 and solvent dichloromethane molecule [40]. The values of k_{ELT} for ZnFP-Py are much larger than those of ZnP-Py. The ELT should occur from the HOMO of the pyrene to the next HOMO of the porphyrin. Electron density at 5-C of the next HOMO of ZnFP-Py is larger than that of ZnP-Py, possibly resulting in an enhancement of electronic communication [54-56] between the porphyrin and the pyrene. Furthermore, the -ΔG of ELT of ZnFP-Py (0.16 eV in toluene, 0.55 eV in dichloromethane, and 0.63 eV in acetonitrile) were larger than those of ZnP-Py (-0.23 eV in toluene, 0.11 eV in dichloromethane, and 0.23 eV in acetonitrile). These findings support the observation that the ELT from the porphyrin S_2 of ZnFP-Py is faster than that of ZnP-Py.

In summary, the S_2 fluorescence of porphyrin is quenched through rapid ELT from the pyrenyl group directly connected at the *meso* position, whereas the S_1 fluorescence cannot be quenched. The CT state relaxes to the S_1 state through rapid CR with a high quantum yield. The ELT from the electron donor to the porphyrin S_2 can occur, suggesting that the photooxidation of various molecules by using porphyrin S_2 state, which has higher oxidative ability than the S_1, is possible.

Table 5. Quantum yields of the relaxation processes and the ELT rate constant of *meso*-pyrenylporphyrins

Compound	Solvent	Φ_{ELT}	Φ_{IC}	$\Phi_{CR\rightarrow S1}$	k_{ELT} / 10^{12} s^{-1}
ZnP-Py	toluene	0.50	0.50	0.92	0.88
	dichloromethane	0.76	0.24	0.81	2.00
	acetonitrile	0.90	0.10	0.82	2.60
ZnFP-Py	hexane	0.58	0.42	0.84	3.30
	toluene	0.61	0.39	0.96	3.10
	dichloromethane	0.87	0.13	0.80	40.00
	acetonitrile	0.85	0.15	0.96	10.00

6. Control of Singlet Oxygen-Generating Activity of Pyrene-Connecting Porphyrin by DNA

6.1. Photodynamic Therapy

The porphyrin photosensitizer is an important drug in PDT. It is a less invasive treatment for cancer, which employs a photosensitizer and visible-light to produce an oxidative stress in cells and ablate cancerous tumors [6-8]. Human tissue has relatively high transparency for visiblelight, especially red light, and visiblelight hardly demonstrates side-effects. The general procedure of PDT is as follows: 1) Administration of photosensitizer (drug), 2) photo-irradiation to photosensitizer, and 3) photosensitized damage of biomacromolecules of cancer.

This photodamage leads to cell death of the tumor via apoptosis or necrosis. Porphyrins have a relatively strong absorption band around the red region. Furthermore, in general, porphyrins can easily generate $^1O_2$, an important reactive oxygen species of PDT, to photosensitize biomolecules oxidation.

Therefore, various porphyrins are clinically used as the PDT photosensitizers. DNA is one of the important target biomacromolecules for PDT [59], and DNA-targeting drugs have been studied [60,61]. It has been reported that $^1O_2$ selectively oxidizes only guanine [62]. The main oxidized product of guanine by $^1O_2$ is 8-oxo-7,8-dihydro-2'-deoxyguanine in both isolated and cellular DNA [63, 64].

6.2. Design of Pyrene-Connecting Porphyrin for the PDT Photosensitizer

The activity control of photosensitized $^1O_2$ generation by DNA should be the initial step to achieve the tailor-made PDT. It has been reported that the photosensitizing alkaloids, berberine and palmatine, can easily bind to form a complex with the DNA through electrostatic interaction, and can generate $^1O_2$ only when the DNA-photosensitizer complex is formed [65-67].

The interaction changes the redox potentials of these photosensitizers to suppress the quenching by intramolecular ELT, resulting in the elongation of the lifetime of the photoexcited state, and making the ENT to molecular oxygen possible [66, 67]. Generated $^1O_2$ effectively oxidizes every guanine residue of DNA [65]. However, these photosensitizers cannot absorb the long wavelength light, the use of which is advantageous for PDT. Thus, the electron donor-connecting porphyrin photosensitizer, which is important for clinical use because of its high absorptivity for the red region (> 630 nm), was designed [5].

The designed photosensitizer is *meso*-(1-pyrenyl)-tris(*N*-methyl-*p*-pyridinium)porphyrin (Py-TMPyP, Figure 13). The MO calculation was performed at the Hartree-Fock 6-31G* level to predict the photophysical property of the porphyrinoid photosensitizer. This calculation showed that the photoexcited state of Py-TMPyP can be deactivated via intramolecular ELT from the pyrene moiety to the porphyrin moiety, forming a CT state. The interaction with DNA predicts a raise in the CT state energy, leading to the recovery of the photochemical activity, such as berberine and palmatine [65-67] and previsouly reported porphyrin [68].

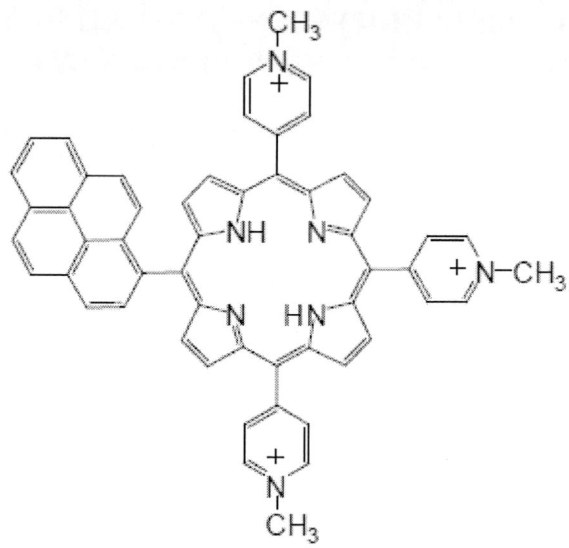

Py-TMPyP

Figure 13. Structure of Py-TMPyP.

6.3. Synthesis of Pyrene-Connecting Porphyrin

The electron donor-connecting porphyrin, *meso*-(1-pyrenyl)-tris(*p*-pyridyl)porphyrin (Py-TPyP), was synthesized from 1-pyrenecarboxaldehyde, 4-pyridinecarboxaldehyde, and pyrrole, according to the literature (Figure 14) [69, 70]. These reagents were dissolved in propionic acid to reflux for 1 hour. After this reaction, sodium acetate trihydrate was dissolved in water and added to the reaction mixture, and stirred to bring the pH to approximately 3.0. The resulting purple precipitate was collected by Büchner filtration and washed with *N,N*-dimethylformamide and methanol. The crude product was purified by column chromatography on silica gel with chloroform-methanol (95:5, v/v). This porphyrin was treated with methyl iodide to obtain Py-TMPyP [71].

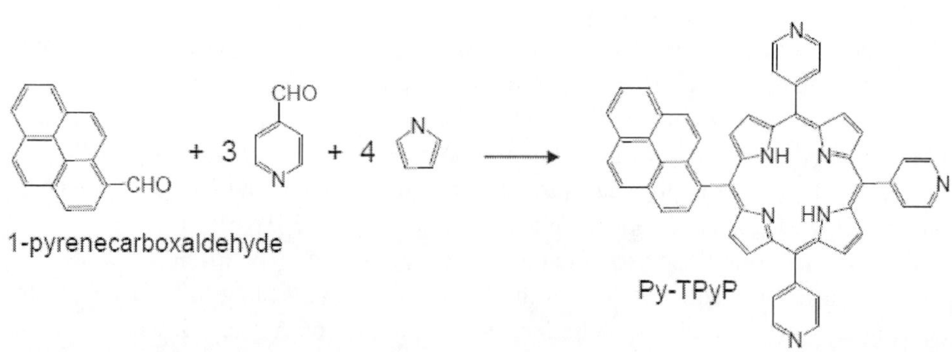

Figure 14. Synthesis of Py-TPyP from 1-pyrenecarboxaldehyde, 4-pyridinecarboxaldehyde, and pyrrole.

6.4. Photochemical Property of Py-TMPyP

The UV-Vis absorption spectrum of Py-TMPyP was red-shifted by the addition of DNA, indicating the binding interaction of Py-TMPyP into the DNA strand. From the CPK model, Py-TMPyP is considered to bind to the DNA major groove. A similar binding form of similar cationic porphyrin has been reported [72]. As the DNA, the synthesized 16-mer oligonucleotides (AATT: $d(AAAATTTTAAAATTTT)_2$ and AGTC: $d(AAGCTTTGCAAAGCTT)_2$ were used. The apparent association constants (K_{app}), which were estimated from the relationship between the absorbance change and the DNA concentration according to the literature [73], are 1.0×10^6 M^{-1} and 8.3×10^5 M^{-1} for AATT and AGTC, respectively. The interaction between Py-TMPyP and AATT is slightly higher than that of AGTC. Py-TMPyP showed almost no fluorescence in an aqueous solution (fluorescence quantum yield, $\Phi_f < 1\times10^{-3}$). The extremely weak fluorescence for free Py-TMPyP stems from an effective quenching of the photoexcited state by the intramolecular ELT from the pyrene moiety to the porphyrin moiety to form the CT state. Note that in the presence of DNA, the fluorescence intensity increased. The apparent values of Φ_f are 0.12 and 0.10 in the presence of 50 μM-base pair (μM-bp) AATT and AGTC, respectively. Owing to the positive charge in the methylpyridinio group of porphyrin, it can interact with anionic DNA. The electron-accepting ability of the porphyrin moiety should be decreased by the electrostatic interaction with anionic DNA, as with the previously reported photosensitizers, berberine and palmatine [65-67]. The DNA-induced decrease (ΔE) in the $-\Delta G$ is about 0.12 eV for palmatine-DNA [74]. In the present study, the decrease in the energy ΔE was estimated to be 0.1-0.37 eV. Because the roughly estimated $-\Delta G$ of Py-TMPyP without DNA is about 0.31 eV, the control of the ELT by DNA is possible from an energetically point of view. Furthermore, the hydrophobic environment of DNA is also unfavorable for the CT state. Thus, the intramolecular ELT was suppressed, resulting in the enhancement of the fluorescence due to the elongation of the S_1 state lifetime.

6.5. Control of Singlet Oxygen Generation by DNA

To evaluate the $^1O_2$ generating activity of the photosensitizer, the near-infrared emission was measured. The typical near-infrared emission spectrum of approximately 1,270 nm, which is assigned to the deactivation of $^1O_2$ to its ground state, was clearly observed during the photoexcitation of Py-TMPyP with DNA, whereas the emission was not observed without DNA (Figure 15). The near-infrared emission was effectively diminished by sodium azide, a physical quencher of $^1O_2$. The quantum yield of $^1O_2$ generation (Φ_Δ) was estimated from the comparison of the emission intensity by Py-TMPyP-DNA and that of methylene blue (0.52 in water) [75]. The apparent values of Φ_Δ by Py-TMPyP-DNA were 0.051 and 0.038 in the presence of 50 μM-bp AATT and AGTC, respectively. Thus, the photosensitized $^1O_2$ generation by Py-TMPyP became possible through the interaction with DNA. Because $^1O_2$ generation occurs near DNA, the generated $^1O_2$ should interact with the DNA strand. AT sequences quench $^1O_2$ through mainly a physical mechanism with the rate constant of 4.1×10^5 $M^{-1}s^{-1}$ [76], whereas guanine can quench $^1O_2$ through a chemical process (guanine oxidation) with a higher rate constant (1.7×10^7 $M^{-1}s^{-1}$) [77]. Therefore, the Φ_Δ may be higher than the estimated values.

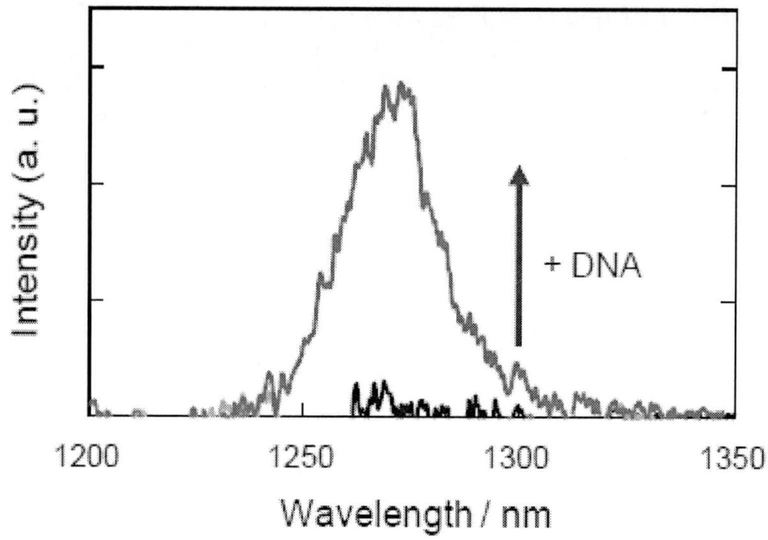

Figure 15. Near-infrared emission spectra of Py-TMPyP with or without DNA.

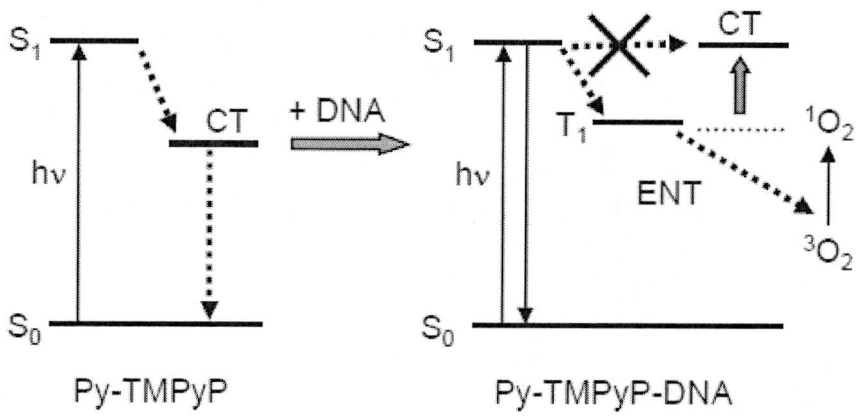

Figure 16. Scheme of the control of $^1O_2$ generation by Py-TMPyP through interaction with DNA.

The S_1 state of Py-TMPyP is effectively quenched through intramolecular ELT from the pyrene moiety (Figure 16). The interaction with DNA suppresses the intramolecular ELT, resulting in an increase in the fluorescence intensity. This suppression of the ELT quenching enhances the intersystem crossing and makes the ENT to molecular oxygen possible. The activity control of porphyrin photosensitizers for PDT through interaction with DNA should provide a possible increase in the selectivity for targeting DNA.

CONCLUSION

Pyrene can act as an ELT and ENT donor against the porphyrin. These photochemical processes can be controlled by the solvent polarity, distance between the donor and the acceptor, and pH of the aqueous solution. Furthermore, ultra-fast ELT from the pyrene to the photoexcited porphyrin was observed in the direct-connecting molecular system, *meso*-pyreneylporphyrin. These photochemical characters of pyrene should be applied for the investigation of fundamental photochemistry and the development of various molecular devices. In addition, the electron transition of pyrene could be used to control the activity of the photosensitizer. The photosensitized $^1O_2$ generation is controlled through the ELT from the pyrene. Medical and medicinal application of this simple molecule, pyrene, might be possible in near future.

REFERENCES

[1] Handbook of Photochemistry, Third Edition, (2006). CRC Press, Florida.

[2] Hirakawa, K. and Segawa, H. (1999). Excitation energy transfer and photo-induced electron transfer in axial bispyrenyl phosphorus porphyrin derivatives: factors governing the competition between energy and electron transfer processes under the existence of intramolecular π-π interaction. *J. Photochem. Photobiol. A: Chem.*, *123*, 67-76.

[3] Hirakawa, K. and Segawa, H. (2010). Acid dissociation of axial hydroxyl group of hydroxy(1-pyrenebutoxy)-phosphorus(V) porphyrin controls the intramolecular excitation energy transfer. *Photochem. Photobiol. Sci., 9*, 704-709.

[4] Hirakawa, K. and Segawa, H. (2010). Effects of hydrogen bonding interaction and solvent polarity on the competition between excitation energy and photo-induced electron transfer processes in hydroxy(1-pyrenebutoxy)phosphorus(V) porphyrin. *J. Photochem. Photobiol. A: Chem., 213*, 73-79.

[5] Hirakawa, K.; Harada, M; Okazaki, S. and Nosaka, Y. (2012). Controlled generation of singlet oxygen by a water-soluble *meso*-pyrenylporphyrin photosensitizer through interaction with DNA. *Chem. Commun.*, *48*, 4770-1772.

[6] Dolmans, D. E. J. G. J.; Fukumura, D. and Jain, R. K. (2003). Photodynamic therapy for cancer. *Nat. Rev. Cancer*, *3*, 380-387.

[7] Castano, A. P.; Mroz, P. and Hamblin, M. R. (2006). Photodynamic therapy and anti-tumor immunity. *Nat. Rev. Cancer,* ***6,*** 535-545.

[8] Wilson, B. C. and Patterson, M. S. (2008). The physics, biophysics and technology of photodynamic therapy. *Phys. Med. Biol.*, *53*, R61-R109.

[9] Pysh, E. S. and Yang, N. C. (1963). Polarographic oxidation potentials of aromatic compounds. *J. Am. Chem. Soc.*, *85*, 2124-2130.

[10] Parker, C. A. and Joyce, T. A. (1966). Triplet formation efficiencies from delayed fluorescence measurements. *J. Chem. Soc. Chem. Commun.*, 234-235.

[11] Stewart, G. M. and Fox, M. A. (1996). Chromophore-labeled dendrons as light harvesting antennae. *J. Am. Chem. Soc.*, *118*, 4354-4360.

[12] Devadoss, C.; Bharathi, P. and Moore, J. S. (1996). Energy transfer in dendritic macromolecules: molecular size effects and the role of an energy gradient. *J. Am. Chem. Soc.*, *118*, 9635-9644.

[13] Wilkinson, F. (1988). Competition between energy and electron transfer, in: Fox, M. A. and Chanon, M. (Eds.), Photoinduced Electron Transfer, Chapter 1.5, Elsevier, pp. 207-227.

[14] Marrese, C. A. and Carrano, C. J. (1983). Synthesis, characterization, and electrochemistry of (5,10,15,20-tetraphenylporphinato) dichlorophosphorus(V) chloride. *Inorg. Chem., 22*, 1858-1862.

[15] Weller, A. (1982). Photoinduced electron transfer in solution. *Z. Phys. Chem. Neue Folge*, *133*, 93-98.

[16] Breton, J. (1988). Vermeglio, A. (Eds.), The photosynthetic bacterial reaction center, structure and dynamics, Plenum Press, New York.

[17] Deisenhofer, J. and Michel, H. (1989). The photosynthetic reaction center from the purple bacterium *Rhodopseudomonas viridis. Angew. Chem. Int. Ed. Engl.*, *28*, 829-847.

[18] Huber, R. (1989). A structural basis of light energy and electron transfer in biology. *Angew. Chem. Int. Ed. Engl.*, *28*, 848-869.

[19] Förster, T. (1948). Zwischenmolekulare energiewanderung und fluoreszenz. *Ann. Physik*, *437*, 55-75.

[20] Marrese C. A. and Carrano, C. J. (1982). The redox chemistry of dichlorophosphorus(V) tetraphenylporphine: isolation and characterization of a stable porphyrin anion radical and evidence for internal electron transfer. *J. Chem. Soc. Chem. Commun.*, 1279-1280.

[21] Fuhrhop, J. H.; Kadish, K. M. and Davis, D. G. (1973). The redox behavior of metallo-octaethylporphyrins. *J. Am. Chem. Soc.*, *95*, 5140-5147.

[22] Gouterman, M.; Sayer, P.; Shankland, E. and Smith, J. P. (1981). Porphyrins. 41. phosphorus mesoporphyrin and phthalocyanine. *Inorg. Chem.*, *20*, 87-92.

[23] Kunimoto, K.; Segawa, H. and Shimidzu, T. (1992). Selective synthesis of unsymmetrical dialkoxyphosphorus(V) tetraphenylporphine derivatives by stepwise substitution of axial position. *Tetrahedron lett.*, *33*, 6327-6330.

[24] Takeuchi, Y.; Hirakawa, K.; Susumu, K. and Segawa, H. (2004). Electrochemical determination of charge transfer direction of center-to-edge phosphorus(V) porphyrin arrays. *72*, 449-451.

[25] Wasielewski, M. R. (1988). Fox, M. A.; Chanon, M. (Eds.) Distance dependencies of electron transfer reactions, *Photoinduced Electron Transfer*, Chapter 1.4, Elsevier.

[26] Wasielewski, M. R. (1992). Photoinduced electron transfer in supramolecular systems for artificial photosynthesis. *Chem. Rev.*, *92*, 435-461.

[27] Kurreck, H. and Huber, M. (1995). Model reactions for photosynthesis-photoinduced charge and energy transfer between covalently linked porphyrin and quinone units. *Angew. Chem. Int. Ed. Engl.*, *34,* 849-866.

[28] Piotrowiak, P. (1999). Photoinduced electron transfer in molecular systems: recent developments. *Chem. Soc. Rev.*, *28,* 143-150.

[29] Bajema, R.; Gouterman, M. and Rose, C. B. (1971). Porphyrins XXIII: fluorescence of the second excited singlet and quasiline structure of zinc tetrabenzporphin. *J. Mol. Spectrosc.*, *39,* 421-431.

[30] Martarano, L. A.; Wong, C. P.; Horrocks, Jr., W. D. and Goncalves, A. M. (1976). Luminescence of yttrium(III), lutetium(III), and thorium(IV) porphyrin complexes. *J. Phys. Chem.*, *80,* 2389-2393.

[31] Tsvirko, M. P.; Stelmakh, G. F.; Pyatosin, V. E.; Solovyov, K. N. and Kachura, T. F. (1980). Fluorescence from upper $\pi\pi^*$ electronic states of lanthanide-porphyrin complexes. *Chem. Phys. Lett.*, *73,* 80-83.

[32] Stelmakh, G. F. and Tsvirko, M. P. (1983). Effect of stored vibrational energy on the fluorescence spectra and quantum yield of metalloporphyrins in solutions. *Opt. Spectrosc.*, *55,* 516-519.

[33] Aaviksoo, J.; Freiberg, A.; Savikhin, S.; Stelmakh, G. F. and Tsvirko, M. P. (1984). Picosecond fluorescence study of the lifetimes of metalloporphyrin S_1 and S_2 states. *Chem. Phys. Lett.*, *111,* 275-278.

[34] Tobita, S.; Kaizu, Y.; Kobayashi, H. and Tanaka, I. (1984). Study of higher excited singlet states of zinc(II)-tetraphenylporphin. *J. Chem. Phys.*, *81,* 2962-2969.

[35] Kurabayashi, Y.; Kikuchi, K.; Kokubun, H.; Kaizu, Y. and Kobayashi, H. (1984). $S_2 \rightarrow S_0$ fluorescence of some metallotetraphenylporphyrins. *J. Phys. Chem.*, *88,* 1308-1310.

[36] Ohno, O.; Kaizu, Y. and Kobayashi, H. (1985). Luminescence of some metalloporphins including the complexes of the IIIb metal group. *J. Chem. Phys.*, *82,* 1779-1787.

[37] Kaizu, Y.; Asano, M. and Kobayashi, H. (1986). Dual fluorescence from two erbium(III) porphyrins. *J. Phys. Chem.*, *90,* 3906-3910.

[38] Kaizu, Y.; Maekawa, H. and Kobayashi, H. (1986). Upper excited-state emission of a covalently linked porphyrin dimer. *J. Phys. Chem.*, *90,* 4234-4238.

[39] Fujitsuka, M.; Cho, D. W.; Solladié, N.; Troiani, V.; Qiu, H. and Majima, T. (2007). S_2- and S_1-states properties of zinc porphyrin polypeptides. *J. Photochem. Photobiol. A: Chem.*, *188*, 346-350.

[40] Chosrowjan, H.; Tanigichi, S.; Okada, T.; Takagi, S.; Arai, T. and Tokumaru, K. (1995). Electron transfer quenching of S_2 state fluorescence of Zn-tetraphenylporphyrin. *Chem. Phys. Lett.*, *242,* 644-649.

[41] Gurzadyan, G. G.; Tran-Thi, T. H. and Gustavsson, T. (1998). Time-resolved fluorescence spectroscopy of high-lying electronic states of Zn-tetraphenylporphyrin. *J. Chem. Phys.*, *108,* 385-388.

[42] Harriman, A.; Hisseler, M.; Trompette, O. and Ziessel, R. (1999). Energy transfer in molecular dyads comprising metalloporphyrin and ruthenium(II) tris(2,2'-bipyridyl) terminals. competition between internal conversion and energy transfer in the upper excited singlet state of the porphyrin. *J. Am. Chem. Soc.*, *121,* 2516-2525.

[43] LeGourriérec, D.; Andersson, M.; Davidsson, J.; Mukhtar, E.; Sun, L. and Hammarström, L. (1999). Photoinduced electron transfer from a higher excited state of a porphyrin in a zinc porphyrin−ruthenium(II) *tris*-bipyridine dyad. *J. Phys. Chem. A*, *103,* 557-559.

[44] Mataga, N.; Chosrowjan, H.; Shibata, Y.; Yoshida, N.; Osuka, A.; Kikuzawa, T. and Okada, T. (2001). First unequivocal observation of the whole bell-shaped energy gap law in intramolecular charge separation from S_2 excited state of directly linked porphyrin−imide dyads and its solvent-polarity dependencies. *J. Am. Chem. Soc.*, *123,* 12422-12423.

[45] Mataga, N; Chosrowjan, H.; Taniguchi, S.; Shibata, Y.; Yoshida, N.; Osuka, A.; Kikuzawa, T. and Okada, T. (2002). Ultrafast charge separation from the S_2 excited state of directly linked porphyrin−Imide dyads: first unequivocal observation of the whole bell-shaped energy-gap law and its solvent dependencies. *J. Phys. Chem. A*, *106,* 12191-12201.

[46] Mataga, N.; Taniguchi, S.; Chosrowjan, H.; Osuka, A. and Yoshida. N. (2003). Ultrafast charge transfer and radiationless relaxations from higher excited state (S_2) of

directly linked Zn-porphyrin (ZP)-acceptor dyads: investigations into fundamental problems of exciplex chemistry. *Chem. Phys.*, *295*, 215-228.

[47] Mataga, N.; Taniguchi, S.; Chosrowjan, H.; Osuka, A. and Kurotobi, K. (2005). Observations of the whole bell-shaped energy gap law in the intra-molecular charge separation (CS) from S_2 state of directly linked Zn–porphyrin–imide dyads: examinations of wider range of energy gap ($-\Delta G_{CS}$) for the CS rates in normal regions. *Chem. Phys. Lett.*, *403*, 163-168.

[48] Fujitsuka, M.; Cho, D. W.; Shiragami, T.; Yasuda, M. and Majima, T. (2006). Intramolecular electron transfer from axial ligand to S_2-excited Sb-tetraphenylporphyrin. *J. Phys. Chem. B*, *110*, 9368-9370.

[49] Fujitsuka, M.; Cho, D. W.; Tojo, S.; Inoue, A.; Shiragami, T.; Yasuda, M. and Majima, T. (2007). Electron transfer from axial ligand to S_1- and S_2-excited phosphorus tetraphenylporphyrin. *J. Phys. Chem. A*, *111*, 10574-579.

[50] Hirakawa, K.; Kawanishi, S.; Hirano, T. and Segawa, H. (2007). Guanine-specific DNA oxidation photosensitized by the tetraphenylporphyrin phosphorus(V) complex via singlet oxygen generation and electron transfer. *J. Photochem. Photobiol. B: Biol.*, *87*, 209-217.

[51] Hirakawa, K.; Saito, K. and Segawa, H. (2009). Anomalously selective quenching of S_2 fluorescence from upper excited state of zinc 5-(1'-pyrenyl)-10,15,20-triphenylporphyrin derivatives through intramolecular charge transfer state, *J. Phys. Chem. A*, *113*, 8852-8856.

[52] Adler, A. D.; Longo, F. R.; Finarelli, J. D.; Goldmacher, J.; Assour, J. and Korsakoff, L. (1967). A simplified synthesis for meso-tetraphenyl porphine. *J. Org. Chem.*, *32,* 476.

[53] Buchler, J. W. (1975). Smith, K. M. (Eds.) Static coordination chemistry of metalloporphyrins, *Porphyrins and metalloporphyrins*, Chapter 5, Elsevier.

[54] Fujii, H. (1994). Characterization of high valent iron porphyrin in catalytic reaction by iron(III) tetrapentafluorophenylporphyrin. *Chem. Lett.,* 1491.

[55] Strachan, J. P.; Gouterman, S.; Seth, J.; Kalsbeck, W. A.; Lindsey, J. S.; Holten, D. and Bocian, D. F. (1997). Effects of orbital ordering on electronic communication in multiporphyrin arrays. *J. Am. Chem. Soc.*, *119,* 11191-11201.

[56] Yang, S. I.; Seth, J.; Balasubramanian, T.; Kim, D.; Lindsey, J. S.; Holten, D. and Bocian, D. F. (1999). Interplay of orbital tuning and linker location in controlling electronic communication in porphyrin arrays. *J. Am. Chem. Soc.*, *121,* 4008-4018.

[57] Asahi, T. and Mataga, N. (1991). Femtosecond-picosecond laser photolysis studies on the dynamics of excited charge-transfer complexes: aromatic hydrocarbon-acid anhydride, -tetracyanoethylene, and -tetracyanoquinodimethane systems in acetonitrile solutions. *J. Phys. Chem.*, *95*, 1956-1963.

[58] Hirata, Y.; Saito, T. and Mataga, N. (1987). Photochemistry of pyrene-trialkylamine systems in polar solvents: hydropyrenyl formation from a new type of long-lived ion pair formed by recombination of free ions. *J. Phys. Chem.*, *91*, 3119-3122.

[59] Hirakawa, K. (2008). DNA damage through photo-induced electron transfer and photosensitized generation of reactive oxygen species. in H. Kimura and A. Suzuki (Eds.), New reserch on DNA damage (pp. 197-219). Nova Science Publishers Inc., New York.

[60] Cló, E.; Snyder, J. W.; Ogilby, P. R. and Gothelf, K. V. (2007). Control and selectivity of photosensitized singlet oxygen production: challenges in complex biological systems. *ChemBioChem*, *8*, 475-481.

[61] Cló, E.; Snyder, J. W.; Voigt, N. V.; Ogilby, P. R. and Gothelf, K. V. (2006). DNA-programmed control of photosensitized singlet oxygen production. *J. Am. Chem. Soc.*, *128*, 4200-4201.

[62] Kawanishi, S.; Hiraku, Y. and Oikawa, S. (2001). Mechanism of guanine-specific DNA damage by oxidative stress and its role in carcinogenesis and aging. *Mutat. Res.*, *488*, 65-76.

[63] Cadet, J.; Douki, T. and Ravanat, J.-L. (2010). Oxidatively generated base damage to cellular DNA. *Free Radic. Biol. Med.*, *49*, 9-21.

[64] Ravanat, J.-L.; Di Mascio, P.; Martinez, G. R.; Medeiros, M. H. and Cadet, J. (2000). Singlet oxygen induces oxidation of cellular DNA. *J. Biol. Chem.*, *275*, 40601-40604.

[65] Hirakawa, K.; Kawanishi, S. and Hirano, T. (2005). The mechanism of guanine specific photo-oxidation in the presence of berberine and palmatine: activation of photosensitized singlet oxygen generation through DNA-binding interaction. *Chem. Res. Toxicol., 18,* 1545-1552.

[66] Hirakawa, K. and Hirano, T. (2008). The microenvironment of DNA switches the activity of singlet oxygen generation photosensitized by berberine and palmatine. *Photochem. Photobiol., 84,* 202-208.

[67] Hirakawa, K.; Hirano, T.; Nishimura, Y.; Arai, T. and Nosaka, Y. (2012). Dynamics of singlet oxygen generation by DNA-binding photosensitizers. *J. Phys. Chem. B*, *116*, 3037-3044.

[68] Hirakawa, K.; Hirano, T.; Nishimura, Y.; Arai, T. and Nosaka, Y. (2011). Control of singlet oxygen generation photosensitized by *meso*-anthrylporphyrin through interaction with DNA. *Photochem. Photobiol.*, *87*, 833-839.

[69] Khan, T. A. and Hriljac, J. A. (1999). Hydrothermal synthesis of microporous materials with the direct incorporation of porphyrin molecules. *Inorg. Chim. Acta*, *294*, 179-182.

[70] Hambright, P. and Fleischer, E. B. (1970). The acid-base equilibria, kinetics of copper ion incorporation, and acid-catalyzed zinc ion displacement from the water-soluble porphyrin α,β,γ,δ-tetra(4-N-methylpyridyl)porphine. *Inorg. Chem.*, *9*, 1757-1761.

[71] Ishikawa, Y.; Yamashita, A. and Uno, T. (2001). Efficient photocleavage of DNA by cationic porphyrin–acridine hybrids with the effective length of diamino alkyl linkage. *Chem. Pharm. Bull.*, *49*, 287-293.

[72] Jin, B.; Lee, H. M.; Lee, Y.-A.; Ko, J. H.; Kim, C. and Kim, S. K. (2005). Simultaneous binding of *meso*-tetrakis(*N*-methylpyridinium-4-yl)porphyrin and 4‘,6-diamidino-2-phenylindole at the minor grooves of poly(dA)·poly(dT) and poly[d(A−T)$_2$]: fluorescence resonance energy transfer between DNA bound drugs. *J. Am. Chem. Soc.*, *127*, 2417-2424.

[73] Pyle, A. M.; Rehman, J. P.; Meshoyrer, R.; Kumar, C. V.; Turro, N. J. and Barton, J. K. (1989). Mixed-ligand complexes of ruthenium(II): factors governing binding to DNA. *J. Am. Chem. Soc.*, *111*, 3051-3058.

[74] Li, J.; Shuang, S. and Dong, C. (2009). Study on the phosphorescence characterizations of palmatine chloride on the solid substrate and its interaction with ctDNA. *Talanta*, *77*, 1043-1049.

[75] Usui, Y. and Kamogawa, K. (1974). Standard system to determine the quantum yield of singlet oxygen formation in aqueous solution. *Photochem. Photobiol.*, *19*, 245-247.

[76] Petroselli, G.; Dántola, M. L.; Cabrerizo, F. M.; Capparelli, A. L.; Lorente, C.; Oliveros, E. and Thomas, A. H. (2008). Oxidation of 2‘-deoxyguanosine 5‘-monophosphate photoinduced by pterin: Type I versus Type II mechanism. *J. Am. Chem. Soc.*, *130*, 3001-3011.

[77] Petroselli, G..; Erra-Balsells, R.; Cabrerizo, F. M.; Lorente, C.; Capparelli, A. L.; Braun, A. M.; Oliveros, E. and Thomas, A. H. (2007). Photosensitization of 2′-deoxyadenosine-5′-monophosphate by pterin. *Org. Biomol. Chem.*, *5*, 2792-2799.

In: Pyrene
Editors: Petr Ruzicka and Tomas Kral

ISBN: 978-1-62417-291-5

Chapter 6

SYNTHESIS OF NOVEL SILYLATED PYRENE UNITS AND THEIR APPLICATION AS A FLUORESCENT LABELING AGENT FOR BIOACTIVE COMPONENTS SUCH AS OLIGODNA AND CHOLESTEROL

Kazuo Shinozuka[1] and Toshiyuki Takeuchi[2]
[1]Graduate School of Engineering, Gunma University, Kiryu Japan
[2]Institute for Molecular and Cellular Regulation, Gunma University, Maebashi Japan

ABSTRACT

The silylation of pyrene causes increments of molar extinction coefficient and fluorescence quantum yield in addition to a bathochromic effect (approximately 10 nm) inboth absorption and emission maxima. In order to take advantage of these unique features, the authors of this study developed several novel silylated pyrene analogs and some bioactive components, such as oligoDNA and cholesterol, labeled with silylated pyrene.

Experimental results show that the silylated pyrene can be utilized as a sensitive fluorescent labeling unit for bioactive materials. A summary of the findings is found in this chapter.

1) The silylation of pyrene was achieved *via* the lithiation of bromopyrene, followed by the reaction with appropriate chlorodimethylsilyl agents possessing a modifiable functional group. Subsequent to silylation, a suitable combination of chemical reactions afforded silylated pyrene derivatives with terminal OH or NH_2 groups. The OH group was then converted to phosphoramidite function which is readily applicable to modern DNA/RNA synthesizers utilizing solid-phase phosphoramidite chemistry
2) The silylated pyrene phosphoramidite was used to label the 5'-terminus of oligoDNA *via* phosphodiester linkage. The resulting oligoDNA exhibited a faint fluorescent

signal so long as the oligomer stayed in a single strand. The signal, however, intensified markedly in the presence of a fully complementary (non-mismatched) DNA strand. Interestingly, one-base mismatch introduction into the complementary DNA quenched the signal intensification. Meanwhile, the silylated pyrene containing the NH_2 group was utilized to label the C2'-position of uridine *via* carbamate linkage. The modified uridine was further derived to the corresponding 3'-*O*-phosphoramidite which was subsequently incorporated into two consecutive positions of oligoDNA. This oligoDNA exhibited a marked excimer fluorescence signal only upon binding to a fully complementary DNA molecule. That signal was, in turn, quenched into a single-stranded form and in a duplex with a mismatched base pair. As can be seen, the silylated pyrene-modified oligoDNAs demonstrated a sequence-discriminating property characterized by a lightening of its fluorescence signal.

3) We also applied the silylated pyrene unit to visualize intracellular cholesterol traffic in macrophages. Thus, silylated pyrene possessing NH_2 group was reacted with a cholenic acid derivative. The resulting silylated pyrene-cholesterol exhibited an enhanced fluorescence quantum yield compared to that of unmodified pyrene in methanol. In living cells, silylated pyrene-cholesterol moves to oil droplets in macrophages yet it remains in the ER membrane due to the acyl-CoA:cholesterol acyltransferase (ACAT) inhibitor (Sandoz 58-035). These results seem to indicate that cholesterol which contains silylated pyrene is very suitable for the visualization of intracellular cholesterol traffic due to its enhanced fluorescence signal and behaves similarly to natural cholesterol.

INTRODUCTION

The fluorescent labeling method has become an important and a widespread method in modern biological research for the detectionof and quantifying of specific biological substances. These include nucleic acids, peptides, saccharides and lipids. Numerous dyes have been developed to fulfill the needs of each substance to be labeled under differing experimental conditions. For example, cyanine dyes such as Cy-2, Cy-3, and Cy-5, and fluorescein derivatives such as Alexa 488^R, rhodamine, *inter alia* are widely employed as labeling agents. Pyrene, a simple polycyclic aromatic hydrocarbon fluorophore, is also commonly used for labeling because of its chemical stability in addition to its distinctive bluish fluorescence. It has also a reasonably high fluorescence quantum yield ($\Phi_f = 0.32$ in cyclohexane), with λ_{em} about 390 nm with [1].

In 1996, Kyusin and Matsumoto described the introduction of silyl function at the 9- and 10-positions of anthracene, which causes an increment of the molar extinction coefficient and fluorescence quantum yield along with a red shift (bathochromic shift) of both the absorption maxima and the emission maxima [2]. Thus, the fluorescent quantum yield of 9,10-ditrimethylsilylanthracene [3,4] was reportedly 0.95 (Figure 1) in cyclohexane at room temperature [2] which is far greater than that of original anthracene ($\Phi_f = 0.32$ in cyclohexane). Later, Maeda and Mizuno showed that the introduction of a trimethylsilyl group into pyrene resulted in the same effects on the absorption and fluorescence spectra [5,6], whose intensities increase under continued silyl substitution.

The most intense signal was found in 1,3,6,8-tetrakis(trimethylsilyl)pyrene (Φ_f> 0.99) [6]. Introduction of silyl groups into a polycyclic aromatic compound causes an increased energy level of HOMO and decrease that of LUMO of the compound and a net overall decrease in energy gaps between these orbitals due to the Si-associated σ(C-Si)-π* and σ(C-Si)*-π* interaction [2,5-8]. These attribute to the observed unique electronic properties of the silylated polycyclic aromatic compounds. The above silylated pyrene derivatives, however, are inapplicable as a fluorescent labeling agent as their trimethylsilyl group itself is inert and the derivatives have no modifiable functional group. Therefore, we attempted to develop a new silylated pyrene which bears a modifiable silyl function that will serve as a versatile labeling agent for biological substances.

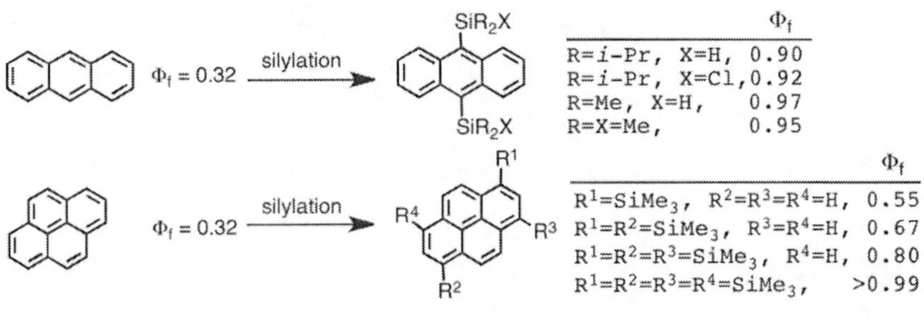

Figure 1. Introduction of TMS group enhances fluorescence of aromatic hydrocarbon [2, 5, 6].

SYNTHESIS OF PYRENE DERIVATIVES WITHMODIFIABLE SILYL FUNCTIONS

The synthesis of silylated pyrene bearing modifiable functional group was achieved by lithiation of halogenopyrene, followed by reaction with the appropriate silylation reagent (Scheme 1). Specifically, treatment of 1-bromopyrene (1) with *n*-BuLi in dry THF at -78 ˚C for 30 min under an argon atmosphere [9] was followed by adding chloro(chloromethyl)dimethylsilane, which in turn produced the corresponding chloromethyldimethylsilylpyrene (2) [10]. Further treatment of compound (2) with sodium acetate in aqueous DMF at 100 ˚C for 24 h produced an acetoxy derivative (3) which was then easily converted to compound 4a bearing a hydroxymethyldimethylsilyl group by the reaction with NaOH in aqueous methanol [10]. In place of chloro (chloromethyl)dimethylsilane as the silylation reagent, chlorodimethyl(vinyl)silane was used and it afforded vinyldimethylsilylpyrene (5) [11]. The compound was then converted to 4b, which bears a hydroxyethyldimethylsilyl group resulting from the hydroboration of (5) with 9-borabicyclo[3.3.1]nonane (9-BBN) and subsequent treatment with EtOH/H_2O_2/NaOH/H_2O [11]. The phosphitylation of 4a and 4b with 2-cyanoethyl *N,N*-diisopropyl chlorophosphoramidite [12,13] gave the corresponding phosphoramidite derivatives (6a, 6b), both of which are applicable for solid-phase oligonucleotide synthesis [14] with slight modification to a standard protocol [10]. On the other hand, the reaction of 4a and 4b with phthalimide in the presence of diethyl azodicarboxylate (DAD) and triphenylphosphine in dioxane resulted in the corresponding pthalimide derivative (7a, 7b).

The subsequent reaction of these compounds with hydrazine hydrate afforded silylated pyrene derivatives having a primary amine function (8a, 8b) [11].

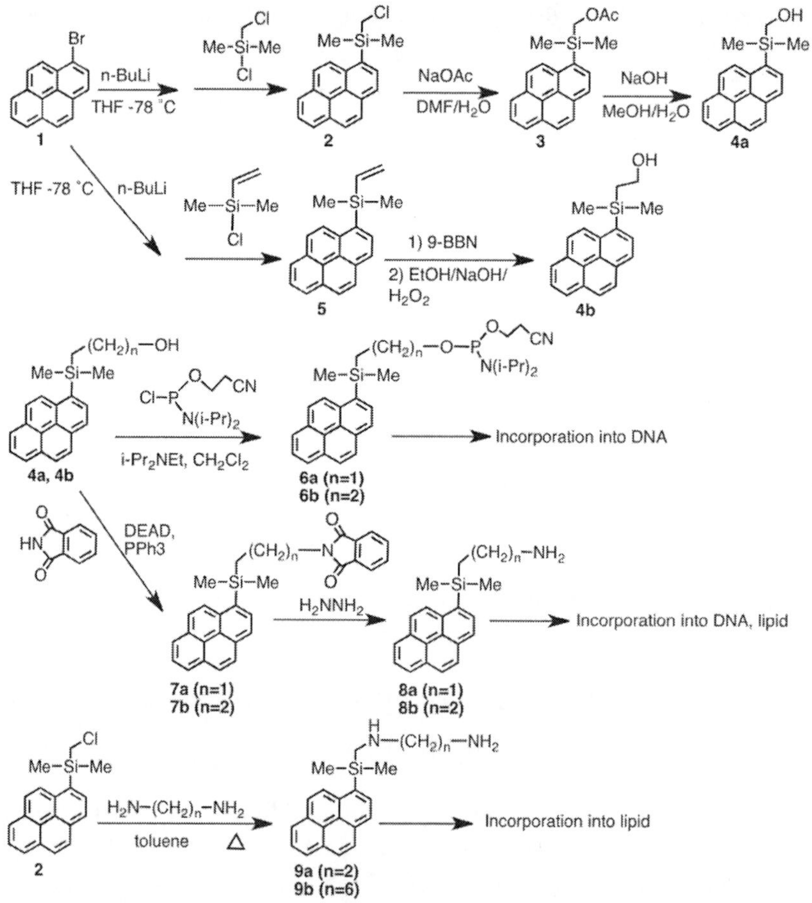

Scheme 1. Synthesis of silylated pyrene derivatives bearing a modifiable functional group.

In addition, an analogous silylated pyrene with a primary amine function (9) was prepared from chloromethyldimethylsilylpyrene (2) with an appropriate alkyldiamine, such as ethylenediamine or hexamethylenediamine [15].

Incorporation of Silylated Pyrene Derivatives Into Biological Substances

First, we prepared an oligonucleotide which bears a silylated pyrene at 5'-terminus through phosphodiester linkage (5'-SiODN). This was produced by the reaction of phosphoramidite reagent 6 with a support-bound oligonucleotide precursor. The precursor had a free 5'-terminal hydroxyl group. This was performed by using an automated DNA synthesizer (Scheme 2). Following the coupling with 6, the support-bound oligomer was subjected to standard deprotection-purification procedures. These included the treatment of conc. ammonium hydroxide (60 ˚C, 12 h) to remove base-revile protecting groups followed by reversed-phase HPLC purification [14]. Under these conditions, the silylated pyrene moiety remained intact [10]. Based on the detailed HPLC analysis of the crude oligonucleotides following the incorporation of 6, however, the higher concentration (0.3 M) of the derivative compared to the concentration used for usual nucleoside phosphoramidite derivatives (0.1 M) seemed to be required to achieve satisfactory coupling yield (>90 %) of 6 [10,16]. Meanwhile, the coupling of aminoethyldimethylsilylpyrene (8b) and 3',5'-*O*-(tetraisopropyldisiloxane-1,3-diyl)uridine (10) with carbonyldiimidazole (CDI) afforded the corresponding ribonucleoside analog bearing silylated pyrene at its C-2' position through carbamate function [17,18] (11b), as depicted in Scheme 3.

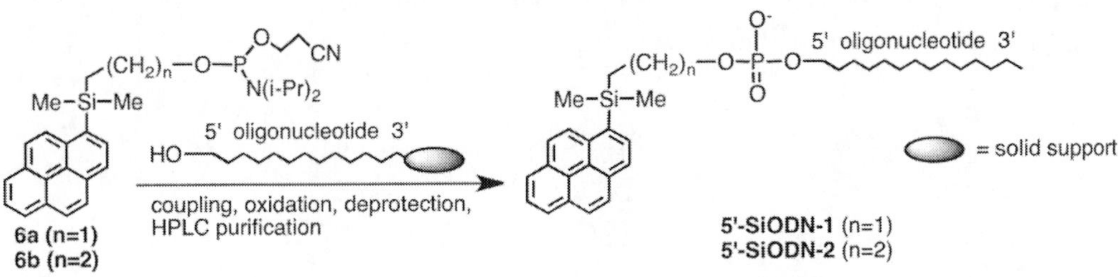

Scheme 2. Incorporation of silylated pyrene into 5'-terminus of oligoDNA.

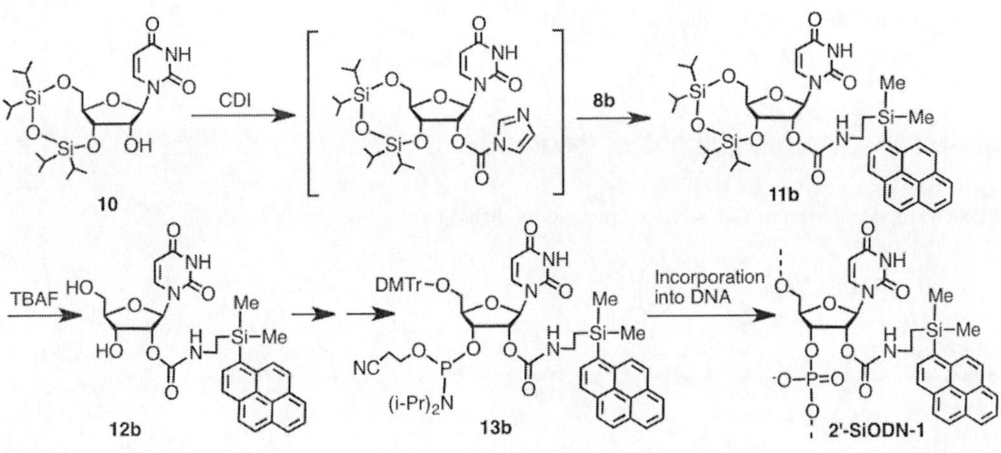

Scheme 3. Incorporation of C-2' silylated pyrene-modified nucleoside residue into oligonucleotide.

After the removal of 3',5'-*O*-protecting group of 11b with tetrabutyla mmonium fluoride (TBAF), the resulting silylated pyrene-modified nucleoside (12b) was reacted with 4,4'-dimethoxytrityl chloride (DMTr-Cl) followed by 2-cyanoethyl *N,N*-diisopropylchlorophosphoramidite to give the corresponding fully protected 3'-*O*-phosphoramoidite derivative (13b). Unlike phosphoramidite reagent 7, compound 13b can be applied to the central as well asthe consecutive positions of oligonucleotide [17,18]. A standard protocol of oligoDNA synthesis [14] with a prolonged coupling period (360 min) for 13b resulted in a coupling yield of approximately 92%. After the assembly, the support-bound oligoDNA was purified by the normal deprotection-purification procedures [14] to give the oligomer (2'-SiODN-1) possessing silylated pyrene-derived nucleoside.

It would be worth to note that compound 8a and 8b were also coupled with C-5 carboxymethy-2'-deoxyuridine (14) [19-21] with the action of an appropriate condensing agent to give the modified deoxyuridine derivatives which have silylated pyrene at C-5 position (15) [22]. The nucleoside was further derived to the corresponding 3'-*O*-phosphoramidite (16) and incorporated into oligonucleotide in the same manner as above (Scheme 4) [22].

Silylated pyrene derivatives were also incorporated into a cholesterol analog. Accordingly, the primary amine-bearing silylated pyrene derivatives (9a, 9b) were reacted with β-hydroxy-Δ^5-cholenic acid under the presence of a condensing agent, 4-(4,6-dimethoxy-1,3,5-triazin-2-yl)4-methylmorpholinium chloride (DMT-MM) [23] in methanol at room temperature for 12h to give the corresponding fluorescent cholesterol analog (17a, 17b) [15], as those are depicted in Scheme 5.

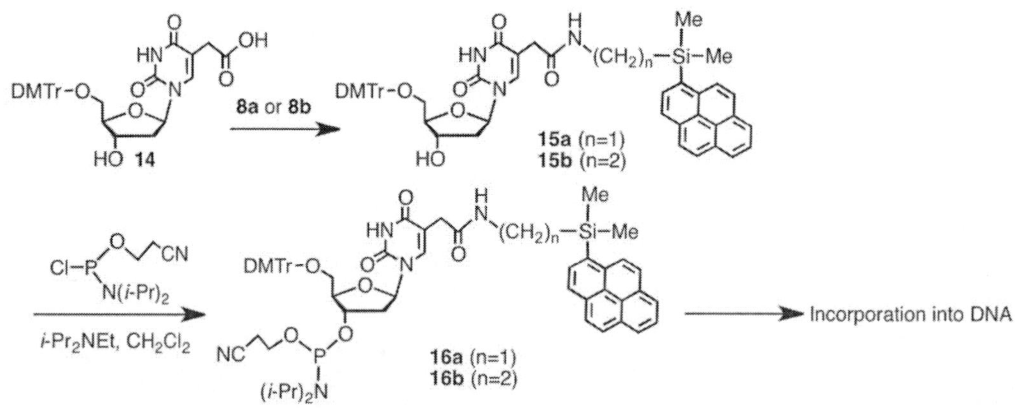

Scheme 4. Incorporation of C-5 silylated pyrene-modified nucleoside into oligonucleotide.

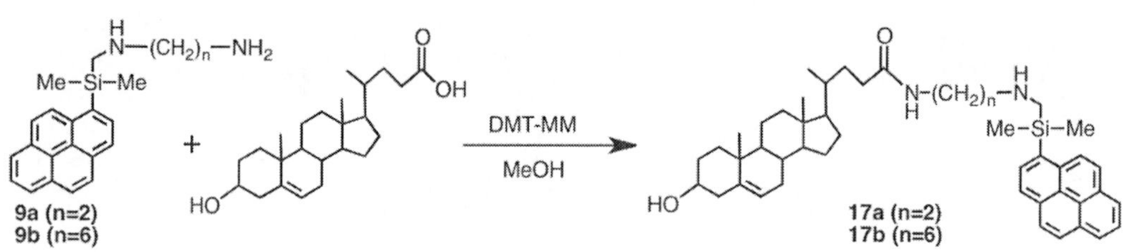

Scheme 5. Synthesis of fluorescent cholesterol analog bearing silylated pyrene.

FLUORESCENT PROPERTIES OF OLIGODNA LABELED WITH SILYLATED PYRENE

Silylated pyrene derivatives which were prepared as described above exhibited an increment of fluorescence quantum yield as well as molar extinction coefficient (ε_{max}). Furthermore, both the absorption maxima (λ_{abs}) and the fluorescent emission maxima (λ_{em}) of silylated pyrene derivatives shifted to longer wavelength (more than 10 nm) compared to those of the unmodified pyrene. For example, the quantum yield of chloromethyldimethylsilylpyrene (2) increased to 0.67 compared with 0.32 of the unmodified pyrene [1,9]. At the same time, the molar extinction coefficient of 2 increased to 8.1 x 10^4 $mol^{-1}dm^3\ cm^{-1}$ (λ_{345}) from the unmodified pyrene's 5.6 x 10^4 $mol^{-1}\ dm^3cm^{-1}$ (λ_{334}) [16], resulting in the fluorescent brightness (Φ_f *times* ε_{max}) of nearly 3 times of unmodified pyrene. Likewise, the quantum yields ofhydroxymethyldimethylsilylpyrene (4a) and hydroxyethyldimethylsilylpyrene (4b) were 0.64 [10] and 0.67 [22], respectively. On the other hand, silylated pyrene incorporated into oligoDNA exhibited somewhat different fluorescent properties from silylated pyrene itself. The oligomer (5'-SiODN-1), which has a complementary sequence to the HIV-1 genome [24] (Figure 2), bearing silylated pyrene at its 5'-terminus through phosphodiester linkage produced nearly imperceptible fluorescence (Φ_f = 0.006) in near physiological solution (100 mM NaOH, pH 7.2) as shown in Figure 3. This result is contrary to previous observations using analogous oligonucleotides bearing a simple pyrene moiety connected at either 5'-phosphate [25] or 5'-hydroxyl group [26] through an alkyl linker. The 5'-SiODN-1's signal, however, intensified by the addition of the fully complementary DNA fragment (cDNA-1) to the solution of 5'-SiODN-1 at 20 °C (Figure 3). UV-melting studies confirmed that upon mixing with cDNA-1 under the above conditions, 5'-SiODN-1 formed a duplex and the melting point (*T*m) of the duplex is 60.6 °C [10].

Moreover, the signal in the duplex disappeared by heating the solution to 90 °C. At this point, the duplex resolved and 5'-SiODN-1 became a single-stranded form because the temperature is well above the melting point. These results indicate that the signal inthe silylated pyrene of 5'-SiODN-1 is effectively quenched while the oligoDNA stays as a single-stranded form. The formation of a duplex, however, appears to cause the marked enhancement of the signal [9].

The effect of single-nucleotide alternations (mismatches) in the target oligoDNA on the fluorescence signal of 5'-SiODN-1 was also examined with cDNA-2 to cDNA-4 possessing single-nucleotide alternation at various positions [10]. The duplex consisting of 5'-SiODN-1 and cDNA-2 produced the most distinct signal among the duplexes containing a mismatched base pair. This was particularly the case when the mismatch-nucleotide was located near the middle of the duplex-forming region. The signal was, however, smaller (ca. 87%) than that of the fully matched duplex (5'-SiODN-1 + cDNA-1).

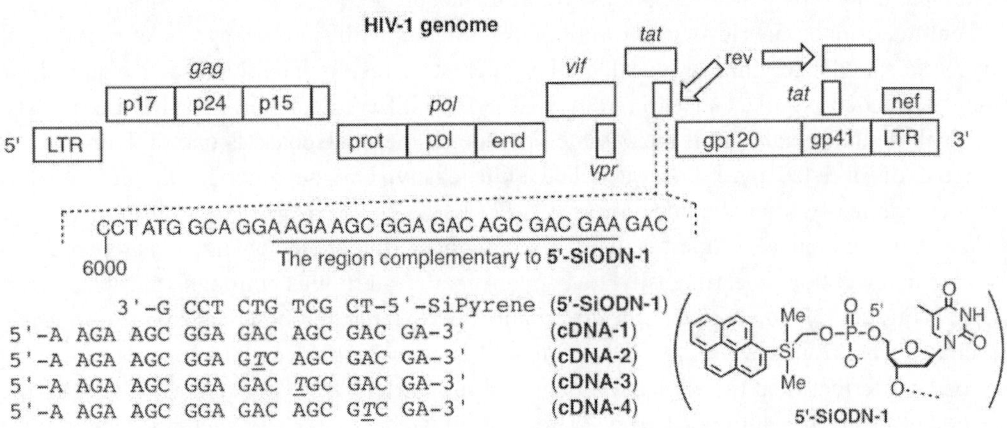

Figure 2. The sequence and structure of 5'-SiODN-1 and its complements. The underlined italicized letters in cDNA-2 to cDNA-4 indicate the corresponding single-nucleotide alternations to the fully matched target DNA (cDNA-1).

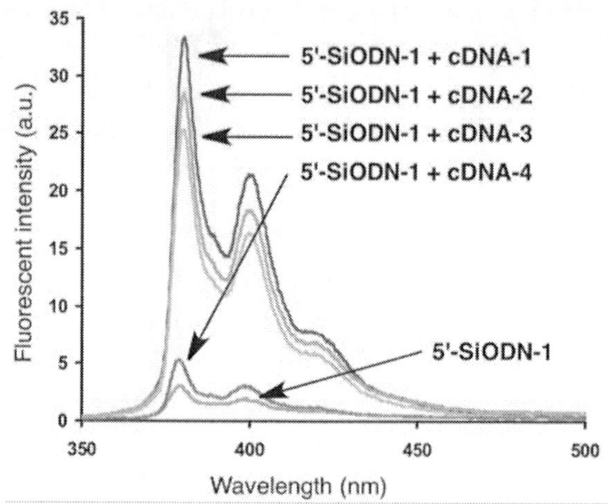

Figure 3. Fluorescence signal of 5'-SiODN-1 with or without complementary DNA. Spectra were recorded in a 10 mM sodium phosphate buffer (pH 7.2) containing 100 mM of NaCl, 0.3 μM of DNA strands using an excitation wavelength of 335 nm.

Meanwhile, the signal decreased as the mismatched position of the targets moved toward 3'-terminus of the duplex-forming region. As a result, the signal strength in the duplex of 5'-SiODN-1 and cDNA-4 was only 14% when compared to that of the fully matched duplex. This is roughly equivalent to that of the single-stranded 5'-SiODN-1. It should be noted that the *T*m value for all duplexes containing mismatched target DNA (cDNA-2 to cDNA-4) was well above (>50 °C) the experimental temperature (20 °C) of the fluorescence measurement study [10].

Therefore, the possibility of imperfect formation of duplex resulting in the single-stranded form of SiODN-1 under the condition should be precluded. Substituting the mismatched T residue in cDNA-4 to other mismatched residues such as G or C also resulted in signal to the same magnitude of the single stranded 5'-SiODN-1 [16]. In conclusion, a rather simple modified oligoDNA labeled with silylated pyrene at its 5'-terminus (5'-SiODN-1) exhibited unique sequence-responding fluorescence properties.

Continuous photometric studies of oligoDNA labeled with silylated pyrene at 5'-terminus through phosphodiester linkage revealed other interesting results. The fluorescence quenching of single-stranded oligoDNA bearing silylated pyrene differs depending on the composition of nucleobase unit neighboring the silylated pyrene. If the nucleobase is one of T or C or G, the signal of silylated pyrene is quenched quite easily. On the other hand, the signal is quenched minimally when the nucleobase is A [27].

We have conducted time-resolved fluorescence studies using newly prepared 2'-deoxymononucleotides bearing silylated pyrene at 5'-terminus through phosphodiester linkage (Figure 4) to explore the mechanism of nucleobase-dependent fluorescence quenching. The results seem to indicate that the quenching is due to the photoinduced electron transfer between the silylated pyrene and the nucleobases G, C and T [28-30]. In the case of G, an electron migrates from G to the silylated pyrene and thereby generating a pyrene anion radical.

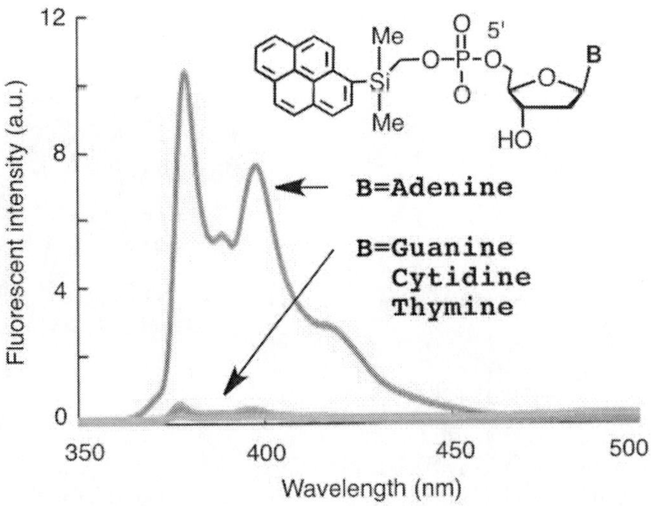

Figure 4. Fluorescence spectra of 5'-mononucleotides bearing silylated pyrene.

In the cases of C and T, on the other hand, an electron migrates from the silylated pyrene to the nucleobase and a pyrene cation radical was generated [28-32]. Thus, the fluorescence signal of the mononucleotide bearing guanine, cytidine or thymine as the nucleobase was effectively quenched (Figure 3). It is worth noting that the introduction of silyl function increases the energy level of HOMO and lowers that of LUMO of pyrene and these likely facilitate the photoinduced electron transfer mentioned above [2,5,7]. Analogously, the fluorescent signal in the uridine derivative (12b) was strongly quenched because uracil moiety (U) of 12b is a demethylated analog of T. An oligonucleotide possessing 12b at two consecutive positions (2'-SiODN-1, Figure 5), thus, exhibited a faint signal ($\Phi_f = 0.04$) as it stays in a single-stranded form in the buffer solution (1.5 μM in 10 mM sodium phosphate containing 100mM NaCl) using excitation wavelength of 314 nm (Figure 6a) [11]. The weak signal found in this stage at approximately 380 to 400 nm was mainly a monomer-state signal of silylated pyrene. It is well known that pyrene forms a characteristic excimer when two pyrene molecules stay in the proximal position and emits a signal circa 460 nm [33,34]. Although two silylated pyrene molecules exist in closed positions, namely the C2'-position of the neighboring nucleoside units, in 2'-SiODN-1, the excimer signal was unobserved in the oligomer at this stage. The excimer-to-monomer fluorescence intensity ratio [35], Iex/Im, was 0.17.

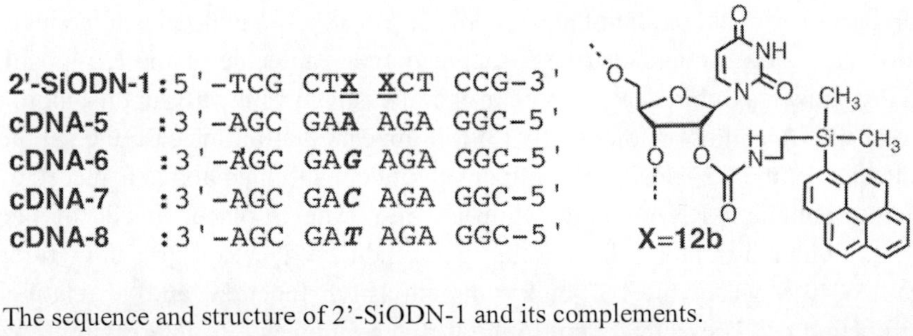

Figure 5. The sequence and structure of 2'-SiODN-1 and its complements.

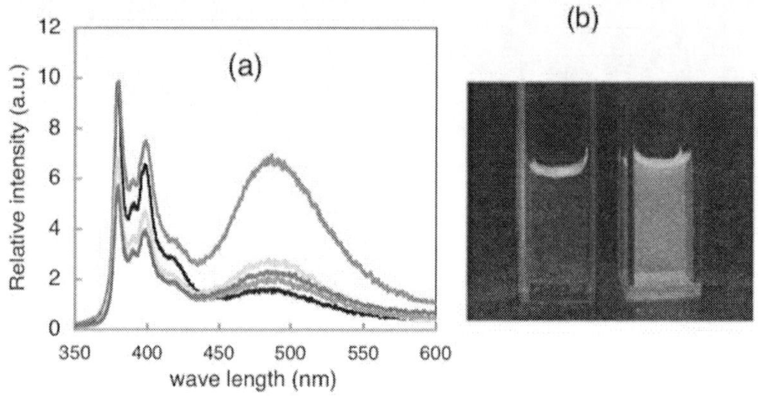

Figure 6.(a); Fluorescent spectra of 2'-SiODN-1. 2'-SiODN-1 alone (black), + cDNA-5 (red), + cDNA-6 (orange), + cDNA-7 (green), + cDNA-8 (blue). (b); Fluorescent signal of 2'-SiODN-1 alone (left) and 2'-SiODN-1 + cDNA-5 (right). (Reproduced with permission from ref. 11. Copyright 2010 Japan Chemical Society).

The addition of a fully complementary DNA fragment (cDNA-5) to the solution of 2'-SiODN-1 at room temperature, however, increased the excimer signal at around 480 nm nearly 5 times of the single-stranded form. Even though the increment of the excimer signal was accomplished within 5 min., the monomer signal was essentially unchanged, as depicted in Figure-6a. The Iex/Im in this stage increased to 0.68. These characteristic changes can readily be observed by the naked eye (Figure 6b). In an oligonucleotide duplex, the C2'-position is faced to a minor-groove of duplex [36]. The results obtained above, therefore, indicate that the silylated pyrene molecules of 2'-SiODN-1 attached to the C2'-position of neighboring uridine residues would not be placed in the closed position while the oligomer remains a flexible single-stranded form in an aqueous solution. Meanwhile, the molecules would be in the proximal position in the duplex consisting of 2'-SiODN-1 and cDNA-5 since they are placed in the narrow minor-groove of the duplex.

The effect of single-nucleotide alternations (mismatches) in the target oligoDNA on the fluorescence signal of 2'-SiODN-1 was also investigated. The fluorescence spectrum of the mixture containing 2'-SiODN-1 and cDNA-6, in which the counter nucleobase to 12b in cDNA-5 was switched from A-residue to G-residue, is depicted in Figure 6a. As can be seen in the spectrum, an increment in neither the monomer-state signal nor the excimer-state signal was observed. The same results were obtained in mixtures containing other mismatched targets (cDNA-7 and cDNA-8). A local perturbation of the duplex structure containing a mismatched base-pair could explain the phenomena [37,38]. An analogous oligonucleotide of modified uridine residues connected to a simple pyrene molecule at the C2'-position by a hydroxymethyl linker exhibited a prominent excimer signal, but only upon binding with a complementary RNA and not with a DNA [39]. It appears that in this case the simple pyrene molecule attached at the C2'-position of oligonucleotide mentioned above is placed at outside of the duplex containing RNA strand while the molecule is placed inside of the duplex (intercalation) containing DNA facilitating photoinduced electron transfer (fluorescence quenching) [39,40]. Meanwhile, a bulky dimethylsilyl function and a relatively rigid carbamate function of 2'-SiODN-1would inhibit pyrene molecule to intercalate to base-pairs in the duplex.

FLUORESCENT PROPERTIES OF CHOLESTEROL ANALOG LABELED WITH SILYLATED PYRENE

The fluorescence spectra of the cholenic acid conjugate of silylated pyrene (17) were measured in degassed methanol, as depicted in Figure 7 [15]. As shown in the Figure, compound 17b bears a long alkyl linker (n=6) between the silylated pyrenyl moiety and the cholenic acid moiety.

This gives off an enhanced fluorescence signal compared to that of the unmodified pyrene. The estimated quantum yield of 17b was 0.65. Contrary to this, the signal of 17a, bearing a short alkyl linker (n=2), was substantially weaker than that of 17b and the estimated quantum yield of 17a was merely 0.03, a quantity even smaller than the unmodified plain pyrene ($\Phi_{\square} = 0.32$).

Because compound 17a was prepared from strongly basic diamine (ethylenediamine), the basicity of the NH-function in 17a could be considerably higher than that of 17b. We suspect that photoinduced electron-transfer could be occurring.

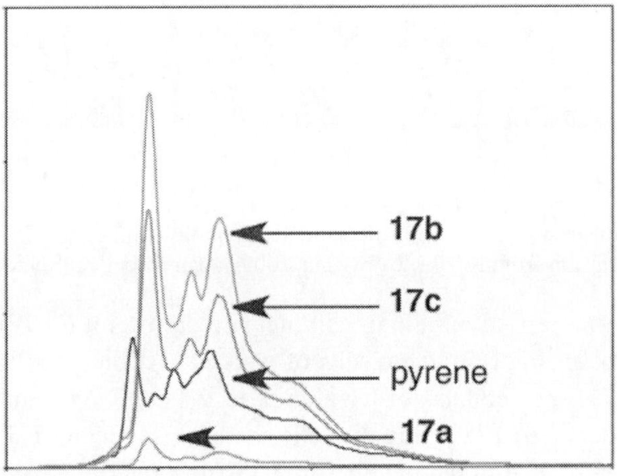

Figure 7. Fluorescent spectra of silylated pyrene conjugated cholesterol analogs 17. (—)Pyrene, (—)17a, (—)17b, and (—)17c in MeOH ($0.1OD_{332}$/mL at room temperature). Excitation wavelength is 332 nm. (Reproduced with permission from ref. 15. Copyright 2009 Japan Chemical Society).

Namely, lone-pair electrons of the NH-function of 17a moved to the pyrene moiety in relatively polar solvent (methanol) under the irradiation of light. To confirm this hypothesis, the NH-function of 17a was acetylated with bromoacetate (Scheme 6) in order to decrease its basicity. The fluorescence signal of the resulting 17c was measured.

As expected, the signal recovered dramatically and the quantum yield of 17c increased to 0.61. Thus, we concluded that the observed low fluorescence of 17a in methanol can be explained by the photoinduced electron-transfer effect described above. This was further confirmed by using an analogous conjugate of cholenic acid and 8b. Fluorescence quantum yield of the conjugate lacking a basic NH-function was 0.67 [41]. In order to assess the utility of 17 as a feasible probe, the uptake and localization of 17a in cultured insulin-secreting pancreatic cells (MIN-6 cells) were monitored using a fluorescence microscope [15].

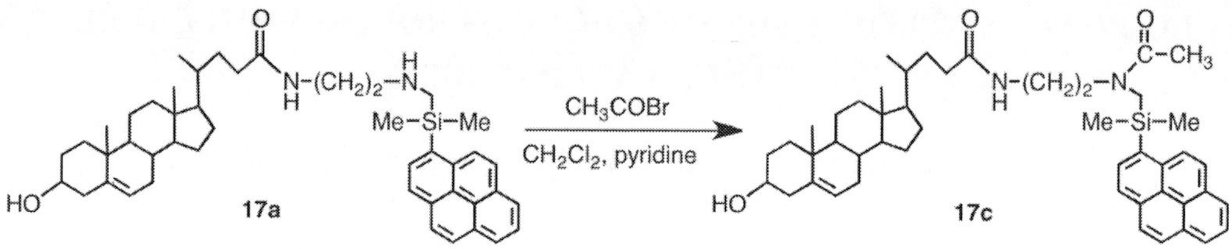

Scheme 6. Acetylation of NH-function of 17a with bromoacetate.

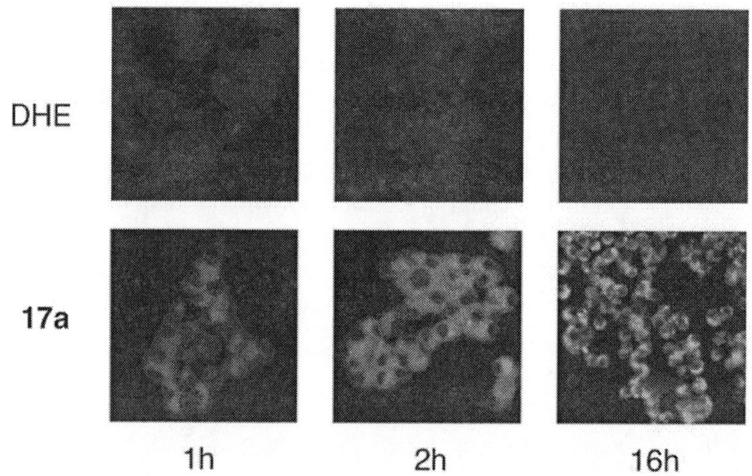

Figure 8. Fluorescent images of MIN-6 cells after incubation with compound 17a and DHE. (Reproduced with permission from ref. 15. Copyright 2009 Japan Chemical Society).

The monitoring was carried out along with dehydroergosterol (DHE), a commonly used fluorescent cholesterol analog [42], the results of which are depicted in Figure 8. The uptake of both DHE and 17a were commenced within 1 to 2 hours. As Figure 8 clearly shows, however, the accumulation of 17a within the cells can be easily identified compared to that of DHE. This could be attributed to the low $\Phi_{\square}$ of DHE (Φ_{ϕ}= 0.04 in ethanol) [43]. In contrast, the localization of 17a in the cell compartments (organelles) remained nearly identical to that of DHE. It should be noted that cholesterol is believed to be localized in the cell's hydrophobic lipid membrane in which the possible photoinduced electron-transfer on 17a would be effectively prohibited. Thus, the new analog behaves nearly the same as the standard analog, yet it is far easier to detect by microscopic observation [44].

APPLICATIONOF THE SILYLATED PYRENE UNIT FOR VISUALIZATION OF INTRACELLULAR CHOLESTEROL TRAFFIC

1) Background of cholesterol visualization

We applied the silylated pyrene unit (cholenic acid conjugate of silylated pyrene 17) to observe intracellular cholesterol traffic in cholesterol-rich cell-types, such as endocrine cells and macrophages. Cholesterol is an essential component of the lipid bilayer membranes in the cells, and is a precursor for steroid hormones and bile acids [45].

In living cells, cholesterol is synthesized in the endoplasmic reticulum (ER) from acetoacetyl-CoA, and is transported to the plasma membrane via the Golgi apparatus. During the transport, cholesterol forms a gradient in composition as it moves from the ER and mitochondria (approximately 5 mol%), Golgi apparatus (10-15 mol%) to the plasma membrane (25-30 mol%) [45-47]. Cholesterol is often concentrated to a specific microdomain - particularly in the plasma membrane – and these concentrations are referred to as rafts (>30 mol%). They seemingly play an essential role in receptor-mediated signal transduction to intracellular functional regions [48]. To visualize cholesterol molecules in live cells, two types of fluorescent cholesterol analogs have been utilized. The first is dehydroergosterol (DHE), a fluorescent analog in plants that behaves as natural cholesterol in animals [42]. DHE has three additional double bonds and an extra methyl group (Figure 9b). The second are the synthetic analogs with fluorescent reporter groups, including (c) NBD, (d) BODIPY, (e) Dansyl, and (f) pyrene (Figure 9). Because DHE exhibits only a faint signal, these fluorescent cholesterol analogs have been produced [43].

Most of them behave in a similar manner to DHE, but some have been found to exhibit different biological behaviors from natural cholesterol [49,50]. This can be likely attributed to a change in hydrophobicity due to ionic charges contained in the reporter unit, or by an inappropriate position connecting wrong connection of a fluorescent reporter unit to a cholesterol molecule.

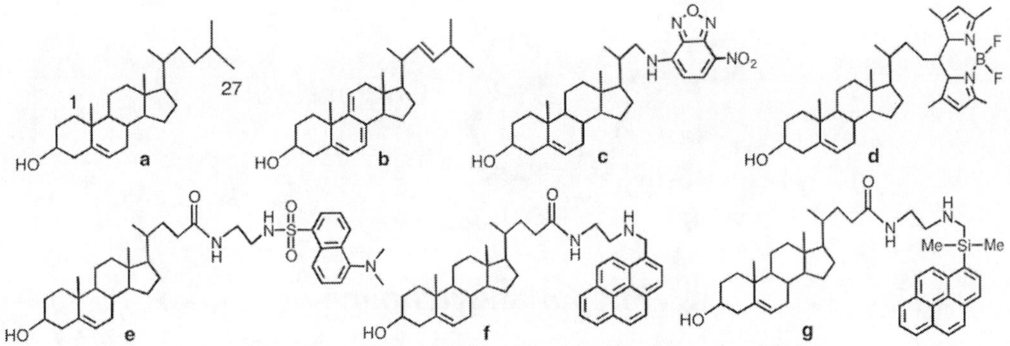

Figure 9. a), cholesterol; b), ergosterol; c), NBD-cholesterol; d), BODIPY-cholesterol; e), Dansyl-cholesterol; f), pyrene-cholesterol; g), pyrene-Si-cholesterol.

A cholesterol molecule is inserted in the phospholipid bilayer membrane with its alkyl side chain deep in the middle of the bilayer and 3β-OH moiety appearing on the bilayer surface to face a phosphatidyl lipid molecule, so that it is positioned in parallel with phospholipid legs [51] (Figure 10). When this polarity is disturbed by ionic charges or a wrongly connected fluorescent unit in a cholesterol analog, it exhibits different intracellular traffic from natural cholesterol [42].

Accordingly, a novel fluorescent analog which has both the property of natural cholesterol behavior yet intense fluorescence is highly desirable. Such an analog could serve as an effective probe to trace the behavior of cholesterol in living cells.

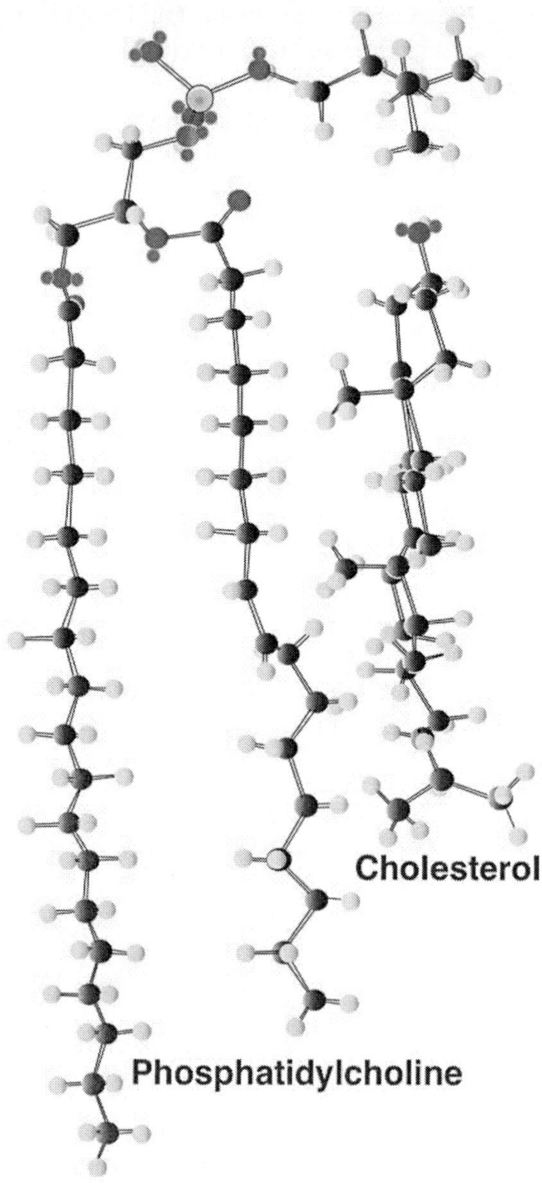

Figure 10. 3D array of phospholipids (phosphatidylcholine) and cholesterol. Black, carbon; yellow, hydrogen; red, oxygen; green, phosphor. Red 1,2, and 3 are tri-carbons from glycerol.

2) Application of silylated pyrene unit to visualize cholesterol molecules

Bearing these features in mind, we attempted to produce a novel fluorescent cholesterol analog using a silylated pyrene unit (17a). This unit demonstrates excellent fluorescent intensity, and is handy in producing newly fluorescence-labeled analogs, as shown in the previous sections. Silylated pyrene bearing an NH_2 group was connected to cholesterol analogs by simple condensation with a cholenic acid derivative. These analogs (pyrene-Si-cholessterol 17a, Figure 9-g) exhibited an enhanced fluorescent quantum yield (greater than 0.6) compared to that of unmodified pyrene (0.32) (Figure 9-f) [15]. Indeed, pyrene-Si-cholesterol 17a exhibited a brighter fluorescence compared with Si-less pyrene-cholesterol (Figure 11, upper panel). Furthermore, the pyrene molecule is immunogenic, and its antibody is commercially available. Thus, we successfully detected cholesterol analogs by fluorescence from silylated pyrene in living cells, and by immunostaining against pyrene in aldehyde-fixed cells (Figure 11, lower panel). By exploiting the advantages of the silylated pyrene-cholesterol 17a, we investigated intracellular cholesterol traffic using two cell-types, pancreatic β-cells and macrophages. Previously, we demonstrated that endocrine secretory granules contain a strikingly high level of cholesterol composition in their membranes [52,53]. In this study, we investigated if exogenously added pyrene-Si-cholesterol 17a accumulates in the secretory granules using a mouse pancreatic β-cell line MIN6. We found that pyrene-Si-cholesterol 17a does indeed accumulate there (Figure 11).

Furthermore, because antibody to pyrene is readily available, we used it to paraformaldehyde-fixed samples in order to visualize pyrene-Si-cholesterol distribution in the cells. Pyrene immunoreactivity is well overlapped with insulin immunostaining, suggesting that cholesterol is enriched at the insulin secretory granules [53]. We then separated the subcelluar organelles of the MIN6 cells to observe the pyrene-Si-cholesterol distribution.

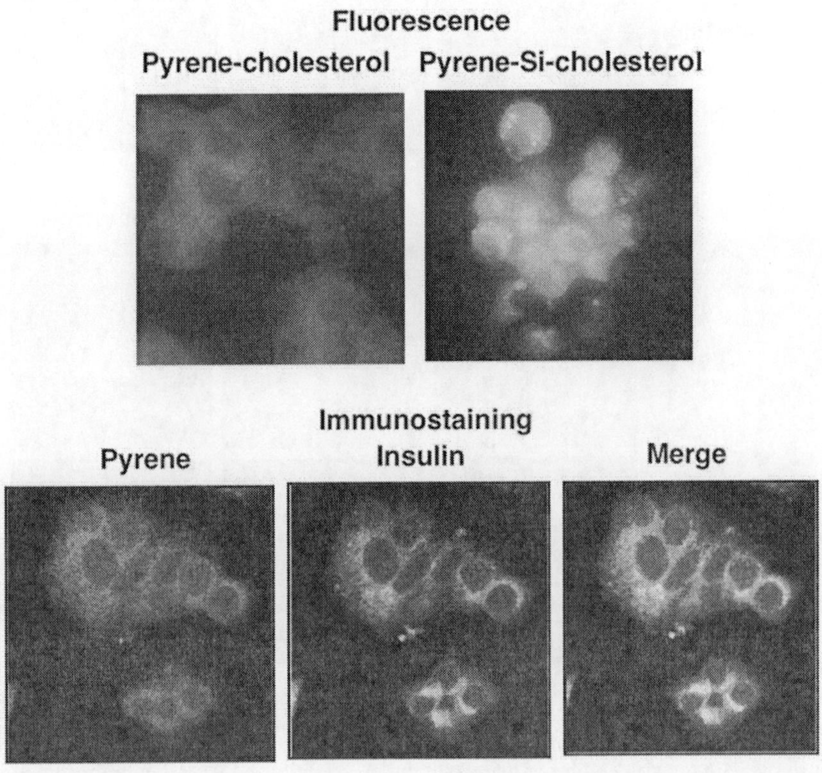

Figure 11. Upper, luminescence from pyrene-cholesterol and pyrene-Si-cholesterol in a mouse pancreatic-β-cell line MIN6. Lower, immunostaining of pyrene and insulin in MIN6 cells.

This was done by sucrose density gradient centrifugation. Pyrene-Si-cholesterol accumulated at the secretory granule fraction, indicated by the insulin peak in a probe dosage-dependent manner. Thus, exogenous cholesterol is transported to the secretory granule membrane. Pyrene-Si-cholesterol was also incorporated in the synaptic-like microvesicle

(SLMV) fraction. Interestingly, synaptic vesicle-residential protein synaptophysin was reported to bind to cholesterol [54], and was also localized to the SLMV fraction in MIN6 cells [52]. Furthermore, secretory granules contain other cholesterol-connecting proteins, secretogranin II and secretogranin III [55,56]. Since these granin proteins easily form aggregates in weakly acidic intragranular milieu where peptide hormones are abundantly found, it would appear that they act as a peptide-hormone carrier into secretory granules by using a high level of cholesterol composition as their anchoring base [57].

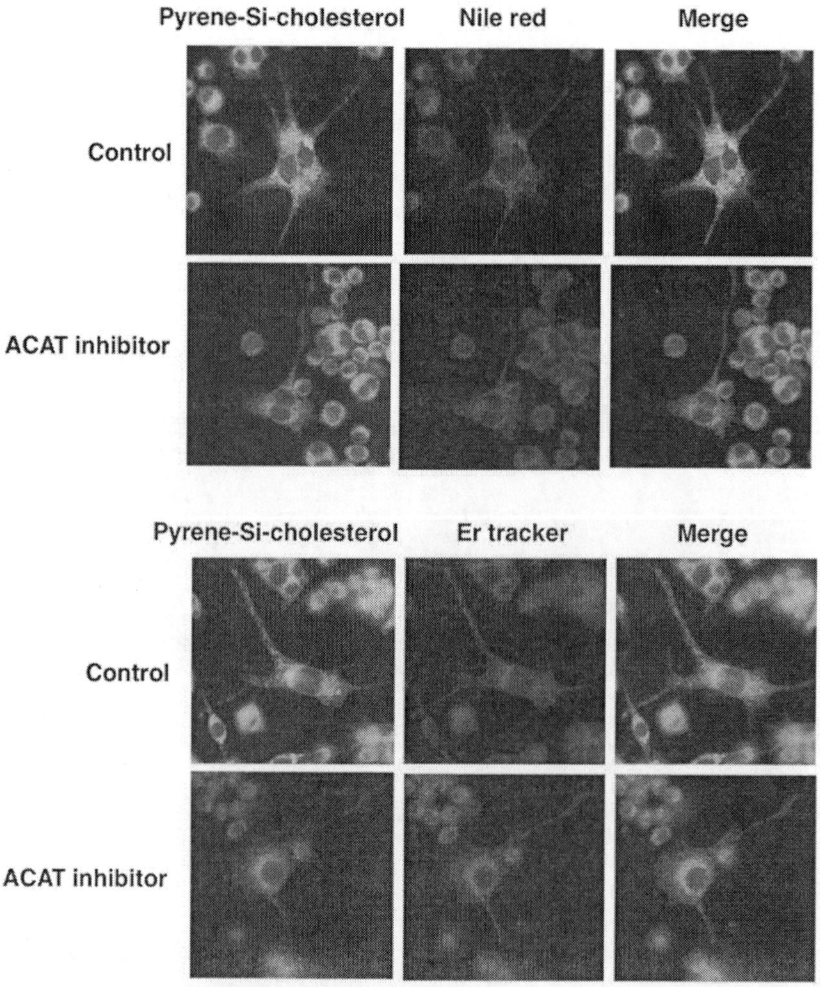

Figure 12. Intracellular traffic of cholesterol from ER to lipid droplets. Without the ACAT inhibitor, pyrene-Si-cholesterol converted to droplets, so that the merged figure shows a blue and red-mixed yellow color in the MIN6 cytosole (upper panel). In the presence of the ACAT inhibitor, merged yellow color changed green, suggesting reduced conversion of pyrene-Si-cholesterol to droplets. The red ER tracker turned green in the merged picture in control (lower panel). In the presence of the ACAT inhibitor, the red ER tracker turned yellow with blue pyrene fluorescence, suggesting that pyrene-Si-cholesterol is blocked to droplets in intracellular traffic by the ACAT inhibitor.

In addition to the cholesterol traffic in pancreatic β-cells, we further examined cholesterol traffic using a murine macrophage cell line RAW264.7. Cholesterol plays a critical role in the formation of atherosclerosis. When cholesterol-rich low density lipoprotein (LDL) is high in the blood stream due to excessive intake of calories, LDL particles accumulate on the artery wall and undergo oxidation. Oxydized LDLs induce inflammatory cell-attracting adhesion molecules and chemokines from endothelial cells, which attract macrophages into the blood vessel intima underneath the endothelial cells. The macrophages ingest oxidized LDLs via highly expressed scavenger receptors. At ingestion, the macrophages become full of lipid droplets and are transformed into foamy cells. Dense populations of foamy cells constitute fatty streaks in the blood vessels, i.e. an initial lesion of atherosclerotic plaque [58]. With this atherosclerosis pathology in mind, we applied pyrene-Si-cholesterol to RAW cells. We found full of blue fluorescence in the cytosol (Figure 12). Because the cholesterol traffic from the ER to droplets is blocked by acyl-CoA: cholesterol acyltransferase (ACAT) inhibitor and cholesterol remains in the ER membrane [59], we looked at the effect of an ACAT inhibitor Sandoz58-035. In the absence of this inhibitor, pyrene fluorescence overlapped with fat-specific Nile-red staining, turning to yellow, which suggests that cholesterol is localized in the lipid droplets. With the inhibitor, however, pyrene fluorescence did not overlap with Nile-red stained fatty droplets (green-colored cells in Figure 12 upper panel), but overlapped well with ER-tracker (yellow-colored cells in Figure 12 lower panel). This suggests that cholesterol traffic to droplets was blocked to the ER by the inhibitor. If so, this indicates that the pyrene-Si-cholesterol probe behaves similarly to natural cholesterol molecules. In conclusion, we found pyrene-Si-cholesterol to be extremely useful for intracellular cholesterol traffic study. It would be worth to note that other pyrene-Si-cholesterol analogs than 17a are also exhibiting some interesting properties as an intracellular cholesterol probe and those will be the subject of future studies.

REFERENCES

[1] Berlman, IB. Handbook of fluorescence spectra of aromatic molecules. Academic Press, N.Y. 1971.

[2] Kyusin, S; Ikarugi, M; Goto, M; Hiratsuka, H; Matsumoto, H. Synthesis and electronic properties of 9,10-disilylanthracenes. *Organometallics,* 1996, 15, 1067-1070.

[3] Cho, H; Harvey, RG. Synthesis of mono- and bis(trimethylsilyl)anthracenes. *J. Org. Chem.,* 1975, 40, 3097-3110.

[4] Roberts, RMG. Structure and dehydrogenation of silylated dihydroanthracenes. A new route to 9- and 9,10-silylated anthracenes. *J. Organomet. Chem.,* 1976, 110, 281-289.

[5] Maeda, H; Inoue, Y; Isida, H; Mizuno, K. UV Absorption and fluorescence properties of pyrene derivatives having trimethylsilyl, trimethylgermyl, and trimethylstannyl groups. *Chem. Lett.,* 2001, 30, 1224-1225.

[6] Maeda, H; Maeda, T; Mizuno, K; Fujimoto, K; Shimizu, H; Inouye, M. Alkynylpyrenes as improved pyrene-based biomolecular probes with the advantages of high fluorescence quantum yields and long absorption/emission wavelengths. *Chem. Eur. J.,* 2006, 12, 824-831.

[7] Maeda, H; Maeda, T; Mizuno, K. Absorption and fluorescence spectroscopic properties of 1- and 1,4-silyl-substituted naphthalene derivatives. *Molecules*, 2012, 17, 5108-5125.

[8] Karatsu, T; Shibata, T; Nishigaki, A; Fukui, K; Kitamura A. π–π and σ-π Interaction in α,ω-dinaphthyl and -dianthryl oligosilanes in solution. *Chem. Lett.*, 2001, 30, 994-995.

[9] Declercq, D; Delbeke, P; De Schryver, FC; Van Meervelt, L; Miller. RD. Ground- and excited-state interaction in di-1-pyrenyl-substituted oligosilanes. *J. Am. Chem. Soc.*, 1993, 115, 5702–5708.

[10] Sekiguchi, T; Ebara, Y; Moriguchi, T: Shinozuka, K. *Bioorg.* Novel sequence-responding fluorescent oligoDNA probe bearing a silylated pyrene molecule. *Bioorg. Med. Chem. Lett.*, 2007, 17, 6883-6886.

[11] Mogi, M; Uddin, MdG; Ichimura, M; Moriguchi, T; Shinozuka, K. Modified oligoDNA having two consecutive silylated-pyrene moieties in minor groove exhibiting an excimer fluorescent signal upon binding to fully-complementary DNA strand. *Chem. Lett.*, 2010, 39, 1254-1255.

[12] Shinha, N.D; Biernat, J; Köster, H. β-Cyanoethyl *N,N*-dialkylamino/ *N*-morpholinomonochloro phosphoamidites, new phosphitylating agents facilitating ease of deprotection and work-up of synthesized oligonucleotides. *Tetrahedron Lett.*, 1983, 24, 5843-5846.

[13] Usman, N; Ogilvie, KK; Jiang, MY; Cedergren, RL. The automated chemical synthesis of long oligoribuncleotides using 2'-*O*-silylated ribonucleoside 3'-*O*-phosphoramidites on a controlled-pore glass support: Synthesis of a 43-nucleotide sequence similar to the 3'-half molecule of an *Escherichia coli* formylmethionine tRNA. *J. Am. Chem. Soc.*, 1987, 109, 7845-7854.

[14] Gait, MJ. Oligonucleotide Synthesis: A Practical Approach, IRL Press, Oxford. 1984.

[15] Moriguchi, T; Hosaka, M; Yoshinari, A; Ozaki, A; Takeuchi, T; Shinozuka, K. Cholesterol analogs labeled with novel silylated fluorescent compounds. *Chem. Lett.*, 2009, 38, 966-967.

[16] T. Sekiguchi, T. Moriguchi, K. Shinozuka, unpublished results.

[17] Seio, K; Wada, T; Sakamoto, K; Yokoyama, S; Sekine, M. Chemical synthesis and properties of conformationally fixed diuridine monophosphates as builiding blocks of the RNA turn motif. *J. Org. Chem.,* 1998, 63, 1429.

[18] Korshun, VA; Stetsenko, DA; Gait, MJ. Novel uridin-2′-yl carbamates: synthesis, incorporation into oligodeoxyribonucleotides, and remarkable fluorescence properties of 2′-pyren-1-ylmethylcarbamate. *J. Chem. Soc., Perkin Trans.1*, 2002, 1092-1104.

[19] Sawai, H; Nakamura, A; Sekiguchi, S; Yumoto, K; Endo, M; Ozaki, H. Efficient synthesis of new 5-substituted uracil nucleosides useful for linker arm incorporation. *J. Chem. Soc., Chem. Commun.*, 1994, 1997-1998.

[20] Ozaki, H; Nakamura, A; Arai, M; Endo, M; Sawai, H. Novel C5-substituted 2'-deoxyuridine derivatives bearing amino-linker arms: synthesis, incorporation into oligodeoxyribonucleotides and their hybridization properties. *Bull. Chem. Soc. Jpn.*, 1995. 68, 1981-1987.

[21] Kohgo, S; Shinozuka, K; Ozaki, H; Sawai, H. Synthesis of a novel 2'-deoxyuridine derivative bearing a cyanomethoxycarbonylmethyl group at C-5 position and its use for versatile post-synthetic functionalization of oligodeoxyribonucleotides. *Tetrahedron Lett.*, 1998, 39, 4067-4070.

[22] Udin, MdG; Moriguchi, T; Ichimura, M; Shinozuka, K. Synthesis and properties of molecular beacon DNA probe bearing novel silylated pyrene derivative. *Key Engineering Materials*, 2011, 497, 47-50.

[23] Kunishima, M; Kawachi, C; Hioki, K; Terao, K; Tani, S. Formation of carboxamides by direct condensation of carboxylic acids and amines in alcohols using a new alcohol- and water-soluble condensing agent: DMT-MM, *Tetrahedron*, 2001, 57, 1551-1558.

[24] Ratner, L.; Haseltine, W.; Patarca, R.; Livak, KJ.; Starcich, B.; Josephs, SF.; Doran, ER.; Rafalski, JA.; Whitehorn, EA.; Baumeister, K.; Ivanoff, L.; Pettway Jr, SR.; Pearson, ML.; Lautenberger, JA.; Papas, TS.; Ghrayeb, J.; Chang, NT.; Gallo, RC.; Wong-Staal, F. Complete nucleotide sequence of the AIDS virus, HTLV-III. *Nature*, 1985, 313, 277-284.

[25] Mann, JS.; Shibata, Y.; Meehan, T. Synthesis and properties of an oligodeoxynucleotide modified with a pyrene derivative at the 5'-phosphate. *Bioconjugate Chem.*, 1992, 3, 554-558.

[26] Yamana, K.; Nunota, K.; Nakano, H.; Sangen, O. A new method for introduction of a pyrene group into a terminal position of an oligonucleotide. *Tetrahedron Lett.*, 1994, 16, 2555-2558.

[27] Shinozuka, K; Sekiguchi, T; Ebara, Y; Moriguchi, T. Synthesis, duplex-forming ability and fluorescent property of oligoDNA bearing silylated fluorophore *Nucleic Acids Res. Sym. Ser.,* 2007, 51, 7-8.

[28] Amann, N; Pandurski, E; Fiebig, T; Wagenknecht, H-A. Electron injection into DNA: Synthesis and spectroscopic properties of pyrenyl-modified oligonucleotides. *Chem. Eur. J.*, 2002, 8, 4877-4883.

[29] Kaden, P; Mayer-Enthart, E; Trifonov, A; Fiebig, T; Wagenknecht, H-A. Real-time spectroscopic and chemical probing, of reductive electron transfer in DNA, *Angew. Chem. Int. Ed.,* 2005, 44, 1636 –1639.

[30] Trifonov, A; Buchvarov, I; Wagenknecht, H-A. Fiebig, T. Real-time observation of hydrogen bond-assisted electron transfer to a DNA base, *Chemical Physics Letters,* 2005, 409, 277–280.

[31] Manoharan, M; Tivel, KL; Zhao, M; Nafisi, K; Netzel, TL. Base-sequence dependence of emission lifetimes for D141018-30-6NA oligomers and duplexes covalently labeled with pyrene: Relative electron-transfer quenching efficiencies of A, G, C, and T nucleosides toward pyrene. *J. Phys. Chem.*, 1995, 99, 17461-17472.

[32] Moriguchi, T; Hattori, M; Sekiguchi, T; Shinozuka, K. Investigation of the mechanism of the fluorescent emission of the silylated pyrene-modified DNA. *Nucleic Acids Res. Sym. Ser.,* 2009, 53, 31-32.

[33] Infelta, PP; Grätzel, M. Statistics of solubilizate distribution and its application to pyrene fluorescence in micellar systems. A concise kinetic model. *J. Chem. Phys.*, 1979, 70, 179-186.

[34] Duhamel, J. New insights in the study of pyrene excimer fluorescence to characterize macromolecules and their supramolecular assemblies in solution. *Langmuir*, 2012, 28, 6527–6538.

[35] Hrdlicka, PJ; Babu, BR; Sørensen, MD; Harrit, N; Wengel, J. Multilabeled pyrene-functionalized 2'-amino-LNA probes for nucleic acid detection in homogenous fluorescence assays. *J. Am. Chem. Soc.*, 2005, 127, 13293-13299.

[36] Saenger, W. Principles of nucleic acid structure. Springer-Verlag, N.Y. 1984.

[37] Modric P; DNA mismatch correction, *Ann. Rev. Biochem*., 1987, 56, 435-446.

[38] Boon, EM; Ceres, DM; Drummond, TG; Hill. MG; Barton, JK. Mutation detection by electrocatalysis at DNA-modified electrodes, *Nature Biotechnology*, 2000, 18, 1096 - 1100.

[39] Mahara, A; Iwase, R; Sakamoto, T; Yamana, K; Yamaoka, T; Murakami, A. Bispyrene-conjugated 2'-*O*-Methyloligonucleotide as a highly specific RNA-recognition probe. *Angew. Chem. Int. Ed.,* 2002, 41, 3648-3650.

[40] Nakamura, M; Fukunaga, Y; Sasa, K; Ohtoshi, Y; Kanaori, K; Hayashi, H; Nakano, H; Yamana, K. Pyrene is highly emissive when attached to the RNA duplex but not to the DNA duplex: the structural basis of this difference. *Nucleic Acids Res*., 2005, 18, 5887-5895.

[41] Yoshinari, A; Shinozuka, K. unpublished result.

[42] Mukherjee, S; Zha, X; Tabas, I; Maxfield, FR. Cholesterol distribution in living cells: Fluorescence imaging using dehydroergosterol as a fluorescent cholesterol analog. *Biophysical J.,* 1998, 75, 1915–1925.

[43] Smutzer, G; Crawford BF; Yeagle PL; Physical properties of the fluorescent sterol probe dehydroergosterol. *Biochim et Biophys Acta*, 1986, 862, 361-371.

[44] Wu, D; Larsen, AL; Færgeman, NJ; Brewer, JR; Sage, D. Selective visualization of fluorescent sterols in *Caenorhabditis elegans* by bleach-rate-based image segmentation. *Traffic*, 2010, 11, 440-454.

[45] Lodish, H; Berk, A; Matsudaira, P; Kaiser, CA; Krieger, M; Scott, MP; Zipursky, SL; Darnell, J. 2004, Molecular Cell Biology:Biomembranes and cell architecture. (5th edition. pp147-157). N.Y., Scientific American Press.

[46] Bretscher, MS; Munro, S. Cholesterol and the Golgi apparatus. *Science*, 1993, 261, 1280-1281.

[47] Simons, K; Ikonen, E. How cells handle cholesterol. *Science*, 2000, 290, 1721-1726.

[48] Brown, DA; Rose, JK; Sorting of GPI-anchored proteins to glycolipid-enriched membrane subdomains during transport to the apical cell surface. *Cell*, 1992, 68, 533-544.

[49] Gimpl, G; Gehrig-Burger, K; Probes for studying cholesterol binding and cell biology. *Steroids*, 2011, 76, 216-231.

[50] Scheidt, H; Muller, P; Herrmann, A; Huster, D. The potential of fluorescent and spin-labeled steroid analogs to mimic natural cholesterol. *J. Biol. Chem.,* 2003, 278, 45563-45569.

[51] Ohvo-Rekilä H, Ramstedt B, Leppimäki P, Slotte JP. Cholesterol interactions with phospholipids in membranes. *Progress Lipid Res*., 2002, 41, 66-97.

[52] Wang, R; Hosaka, M; Han, L; Yokota-Hashimoto, H.; Suda, M.; Mitsushima, D; Torii, S; Takeuchi, T. Molecular probes for sensing the cholesterol composition of subcellular organelle membranes. *Biochim. Biophys. Acta*, 2006, 1761, 1169-1181.

[53] Tsuchiya, M; Hosaka, M; Moriguchi, T; Zhang, S-J; Suda, M; Hashimoto-Yokota, H; Shinozuka, K; Takeuchi, T. Cholesterol biosynthetic intermediates enhance regulated insulin secretion and secretory granule formation in pancreatic β-cells. *Endocrinology*, 2010, 151, 4705-4716.

[54] Thiele, C; Hannah, MJ; Fahrenholz, F; Huttner, WB. Cholesterol binds to synaptophysin and is required for biogenesis of synaptic vesicles. *Nat Cell Biol.*, 2000. 2, 42-49.

[55] Hosaka, M; Suda, M; Sakai, Y.; Izumi, T; Watanabe, T; Takeuchi, T. Sercretogranin III binds to cholesterol in the secretory granule membrane as an adapter for chromogranin A. *J. Biol. Chem.*, 2004, 279, 3627-3634.

[56] Hotta, K; Hosaka, M; Tanabe, A; Takeuchi, T. Secretogranin II binds to secretogranin III and forms secretory granules with orexin and NPY. *J Endocrinol.*, 2009, 202, 111-121.

[57] Takeuchi, T; Hosaka, M. Sorting mechanism of peptide hormones and biogenesis mechanism of secretory granules by secretogranin III, a cholesterol-binding protein, in endocrine cells. *Curr. Diabetes Rev.,* 2008, 4, 31-38.

[58] Ghosh, S. Macrophage cholesterol homeostasis and metabolic diseases: critical role of cholesteryl ester mobilization. *Expert Rev. Cardiovasc. Ther.*, 2011, 9, 329-340.

[59] Khelef, N; Soe, TT; Quehenberger, O; Beatini, N; Tabas, I; Maxfield, FR. Enrichment of acyl coenzyme A:cholesterol *O*-acyltransferase near *trans-Golgi* network and endocytic recycling compartment. *Arterioscler. Thromb. Vasc. Biol.,* 2000, 20, 1769-1776.

INDEX

C

D

E

F

G

N

O

Q

R

S

T

U

V

W

Y

Z